Emergency Preparedness and Disaster Recovery in School Libraries

Emergency Preparedness and Disaster Recovery in School Libraries

Creating a Safe Haven

Christie Kaaland

In consultation with and photographs by William Lokey, former FEMA Federal Coordinating Officer

LIBRARIES UNLIMITED

AN IMPRINT OF ABC-CLIO, LLC
Santa Barbara, California • Denver, Colorado • Oxford, England

Library of Congress Cataloging-in-Publication Data

Kaaland, Christie.
Emergency preparedness and disaster recovery in school libraries : creating a safe haven / Christie Kaaland, in consultation with and photographs by William Lokey, former FEMA Federal Coordinating Officer.
pages cm
Includes bibliographical references and index.
ISBN 978–1–61069–729–3 (paperback) — ISBN 978–1–61069–730–9 (ebook) 1. School libraries—Safety measures. 2. School libraries—Security measures. 3. Emergency management—United States—Planning. 4. Disaster relief—United States. I. Lokey, William M. II. Title.
Z679.7.K33 2015
025.8′2—dc23 2014027534

ISBN: 978–1–61069–729–3
EISBN: 978–1–61069–730–9

19 18 17 16 15 1 2 3 4 5

This book is also available on the World Wide Web as an eBook.
Visit www.abc-clio.com for details.

Libraries Unlimited
An Imprint of ABC-CLIO, LLC

ABC-CLIO, LLC
130 Cremona Drive, P.O. Box 1911
Santa Barbara, California 93116-1911

This book is printed on acid-free paper ∞

Manufactured in the United States of America

To my partner, Bud York, researcher extraordinaire, who scoured the news daily for pertinent material and who helped me seek answers to unanswerable questions. His companionship made this project possible. And to Luke and Natalie, and the dedicated educators who work tirelessly to keep them safe at school.

Contents

Preface

My interest in emergency preparedness and recovery began in the teacher certification graduate course I teach at Antioch University Seattle, a class I began teaching shortly after the shooting at Columbine High School. In this course my students and I discussed what we could do—what was within our capacity as teachers—to keep students safe. It was part of a health, fitness, and safety course, but each year the topic of safety took on greater and greater importance.

While working on this book not a single day went by without a pertinent headline news story. Daily reports of natural disasters flooded the news: hurricanes, tornadoes, avalanches, landslides, and urban wildfires. Heartbreaking stories of stabbings and shootings occurred all too frequently in our schools.

In the course of this writing, I made the decision to exclude names of perpetrators involved in the acts of violence in schools, referenced herein. That decision came about with the personal intent of focusing on the victims of violence and their healing process, giving minimal print attention to those who have caused that grief.

While many schools have experienced great tragedy—whether at the hand of Mother Nature or human folly—remarkable stories of recovery and reclamation provide models for a nation in need. One such model is Columbine High School. The arc of growth that is Jeffco Public Schools today stands as a testament to the power of resilience: Out of experiencing the thinkable, Jeffco's programs stand as a national model for safety that includes improved academic standings, phenomenal graduation rates, and an established culture of respect and caring.

I clearly remember standing outside the portable door at Green Lake Elementary talking to my grandchildren's teacher, Kate Sipe, who said to me, "Sandy Hook changed everything." Following Sandy Hook, I wanted answers; we all did. It was the shooting at Sandy Hook Elementary that prompted me to write this book. But I also feel strongly that instead of delving into their stories, reopening their wounds, personalizing their evidence, or demanding answers as to why such atrocities occur, we as educators must each reflect and ask, "What can I do to help?" and remember the names of Sandy Hook and Columbine not as places of tragedy but of stories of power, resilience, and community.

The greater focus on school violence in this book is intentional. Little can be done to prevent an accidental chemical spill from a truck driving by the school, but when discussing human-caused disasters I placed the greatest emphasis on school violence because it is one disaster we can work to prevent and take action against. We can all "look under every rock" for breaches in security and improved safety measures. We can all contribute to the school's culture of respect and caring.

A final section of this book is titled "Books that Heal." Perhaps it is optimistic to hope for when faced with recovering from a major disaster or crisis, but I believe in the strength of story as a healing agent and the power of books to provide children, particularly young children, with identifiable characters and characters as role models, or simply a place to escape, when they are needed most.

Finally, librarians every day are faced with new challenges and added expectations, so let the information, challenges, ideas, and suggestions presented in this book serve not to add to the workload that already exists but to inform and offer an invitation to increase awareness of ways to help keep kids safer.

Acknowledgments

I want to acknowledge the contributions of many colleagues whose efforts, both knowingly and unknowingly, contributed to this book:

Two consecutive Deans of the School of Education, Antioch University Seattle, Dr. Ed Mikel and Dr. Kelly Vancil, who supported me through this process and who daily model leadership in authentic and remarkable ways.

Helen R. Adams, who provided guidance and encouragement and helped me understand "the process."

Dr. Deborah Levitov, whose professional friendship and editorial advice I treasure.

Sharon Coatney, ABC-CLIO Acquisitions Editor, for her patience, encouragement, and guidance.

Deb Kachel, my school library advocacy partner for a decade.

Kate Sipe (teacher extraordinaire) and the fruitful ideas of her third-, fourth-, and fifth-grade students.

John McDonald, for taking time to share stories of resilience from Jeffco Public Schools and for the contributions he has made nationwide in helping keep schoolkids safe.

Stephanie Ferrell and Laura Pearle, for sharing their inspiring stories of recovery following disasters impacting school libraries.

And, finally, Bill Lokey, who ensured accuracy and whose expertise in emergency management and experience in disaster recovery have provided relief to thousands.

Chapter 1

Library Leadership during Emergency Preparedness and Disaster Recovery

ONE MORE LEADERSHIP ROLE FOR LIBRARIANS? . . . *REALLY?!*

Why would a book for school librarians about emergency preparedness, school violence prevention, and disaster recovery start with a chapter on leadership?

Imagine a school in which all administrators and parents point to the school librarian as the most important person in the school, the last position to be cut when budgets require such. Where the librarian's leadership role is so valued that the school would be in chaos without him or her; and in the case of a school emergency, one of the most critical contributors to the students' overall safety rests in the hands of the school's information, research, and communication specialist.

Whether it is recognized or not, the librarian—by the very nature of his or her position—may well be the individual best positioned to assume a major role in a school's emergency preparedness and disaster recovery process.

Today's librarians are being asked to take on more roles of leadership than ever before, not just in the school but also at the broader school district and community level. With each of these added roles, librarians have an opportunity to play a key leadership role. The connections librarians make within the school, in the district, and with community members can have a positive impact in critical moments during an emergency or other disaster.

The librarian's leadership in disaster planning and emergency preparedness provides opportunities for placing the school library at the heart of the school and community when it is needed most: in the face of a disaster or crisis. Further, after a community experiences a disaster, the library can provide needed assistance in recovery, assuming recovery roles and providing needed services, many of which cannot be found elsewhere in the school.

PEDAGOGICAL THEORETICAL UNDERPINNINGS

In foundational teacher training coursework, teachers are introduced to an important theory on human motivation proposed by Abraham Maslow in his educational classic *Hierarchy of Needs: A Theory of Human Motivation*. This theory proposes that human needs must be met hierarchically at each developmental level before one advances to higher levels. Human needs such as to be loved, socialization, and self-actualization (hierarchically the highest level) cannot be achieved if prior basic needs have not been met. Subsequent to the fundamental physical needs of breathing, food, water, and sleep, the psychological need for *safety* and *security* dominates motivation and behavior.

Thus, according to Maslow's theory, students may be incapable of functioning well in an educational environment unless their psychological needs for security and safety are met. Maslow's theory was developed more than 50 years ago and was groundbreaking at the time, and it has been accepted as pedagogically sound. Today more than ever, his theory reinforces the importance of a well-established, strong emergency preparedness plan in which students clearly understand their roles and the roles of those in charge and have confidence in recognized, familiar adult leaders in the school who help provide that sense of safety. Throughout this book, librarians are urged to be those leaders.

But let's face it, librarians can't do everything or be everywhere, even if they would like to. Therefore, take the suggestions, ideas, and leadership urgings within the pages of this book as just that: suggestions and ideas. Do not think that you will need to run emergency preparedness or disaster recovery for an entire school. Rather, find the place or places within that framework where you can make a difference in planning, preparing, and recovering from natural or human-caused disasters.

Librarians: What Role Best Matches Your Talents?

What skills must librarians possess to best take on a leadership role in helping provide the safest possible environment for students in our schools? The following list provides a few key roles and contributions to consider (they are also discussed briefly elsewhere in the chapter):

- *Communication specialist*
- *Collaboration specialist*
- *Technology specialist*
- *Community organizer:*
  - *Conducting hazard hunts*
  - *Neighborhood risks*
- *Reunification procedure*
- *Kits: supply and resupply kits*
- *Equipment management and inventory*
- *Emergency planning and safety curriculum*
- *Drills: scheduling, conducting, evaluating, revising*
- *Time management: the essence of critical time*
- *Staff training planning*
- *Information specialist*

A NATURAL FIT: LIBRARIAN LEADERSHIP IN EMERGENCY PREPAREDNESS AND RECOVERY

Finding the Appropriate Role of Leadership

In 2012 in her role as AASL (American Association of School Librarians) president, Nancy Everhart conducted a national tour of school library programs as she sought a common thread of attributes of outstanding programs. Her "Vision Tour" sought to find those attributes frequently shared among outstanding school libraries.

In her summary of those visits, she discussed school library programs in which leadership is a core value and the librarian's most critical role. She subsequently challenged *all* school librarians to assume greater roles of leadership.

Everhart describes these leaders as follows: "They were also leaders by involvement on decision-making teams and serving as innovators. It was observed that most were change agents who had their own vision for their program. They knew where they wanted to take it and didn't wait around for others to validate them, often asking forgiveness rather than permission" (Everhart, 2013).

Whatever role(s) the school librarian assumes with regard to school safety and emergency preparedness, it may be necessary while serving on decision-making teams to serve also as a change agent for the school, holding the vision of students' greater safety, security, and well-being at the heart of every decision.

Everhart also notes other attributes of outstanding school library programs, such as those led by librarians involved in *outreach, policy-making, collaboration, professional development,* and *building relationships,* all key attributes for successful emergency preparedness and disaster recovery.

The following list of roles, which is by no means comprehensive, provides suggested leadership for the school librarians related to emergency preparedness and safe school practices.

1. Communication Specialist

Throughout this book, you will find a repeated theme that is *based on multiple reminders from experts in emergency management and first responders recounting experiences in the field* regarding the most critical element during and after a disaster: **communication**.

An entire chapter (Chapter 4, "Communication") devoted to communication provides key strategies for efficient communication and the critical roles the school librarian can play related to communication before, during, and after a disaster.

PERSONAL STORY: The Glacial Philosophy of Preparedness

Since my first days as an emergency manager, I have often been asked about how to get the leadership in a school district, local government, or other organization interested in disaster preparedness. It seems that unless there are a lot of potential disaster events in an area, or a recent disaster has occurred somewhere in the world and gotten a lot of media attention, preparedness is often a low priority for schools.

Even after catastrophic events, such as Hurricane Katrina, national surveys show that only a small percentage of the general population is truly prepared to take care of themselves should a disaster occur and affect them. The attitude often seems to be: "It won't happen here, but if it does it won't affect me. If it does affect me, it won't be that bad. If it is that bad, there is nothing I could do about it anyway. So why prepare?"

My strategy to deal with this developed during my first career working in the polar regions supporting scientific research expeditions. I call it the glacial philosophy of preparedness: "Any force, no matter how small, can move any object, no matter how large, if it is consistently applied for a long enough time."

—Bill Lokey

2. Collaboration Is Key

Of all the roles educators assume, librarians are probably the most well versed in the role of collaborator. It is at the heart of library planning and teaching and within the scope of the librarian's curriculum delivery. Methods for librarian-teacher collaboration range from informal planning while crossing paths in the hall with a classroom teacher to formal student-release fully scheduled planning days. It is difficult for librarians to deliver authentic, content-drive curriculum without some collaborative planning with classroom teachers.

In disaster planning also, collaboration is critical. The broader the range of team members for disaster planning and the greater the commitment of those team members, the more prepared the school will be if disaster strikes.

Within the context of collaboration, the librarian may take one of several roles. For example, as a member of a disaster preparedness committee, the librarian may simply provide committee members with a collaboration framework for decision making.

As a professional collaborator, the librarian is able to perform magical acts of collaboration far more complex than simple information distribution, as the following example illustrates:

> In a small rural community in southwestern Washington when excessive rain brought severe flooding to a local community site, many educators and community members had no idea how to seek reparation assistance from the state or federal government. Holding a critical role on the school's emergency planning committee, the librarian was consulted for this information. After contacting the necessary offices of the federal government Federal Emergency Management Agency (FEMA) and the state's emergency relief department, the librarian expeditiously arranged contacts between community leaders and these organizations, thus helping expedite the rocky road to recovery.

Information Power

While it is unlikely the librarian will be designated as incident commander (IC) for or during a disaster, simply being part of emergency leadership is invaluable. At Sandy Hook Elementary, the school principal was the designated incident commander. Tragically, Principal Dawn Hochsprung, the one to whom members of the school community would have turned as incident commander, was one of the first to lose her life. Simultaneously, the person to whom people would have sought assistance with the complexities surrounding mental health services in recovering from this tragedy, the school psychologist, was also killed.

When emergencies strike, the school needs all hands on deck, and countless stories of heroic deeds from the school's adult population during critical incidents have been recorded. But when a member of the school community has taken it upon him- or herself to be well informed regarding best practices in emergency response prior *to an incident, he or she may be able to step into any number of critical roles at a moment's notice, armed with the most valuable emergency preparedness tool: a depth of knowledge and a clear understanding of what to expect and what to do.*

3. Technology Specialist

With the democratization of devices that provide constantly changing technologies and channels for information delivered rapidly and comprehensively, there arise multiple roles for the librarian as the school's information specialist to assist with emergency planning and disaster recovery.

Sprinkled throughout this book, particularly Chapter 4, "Communication," are multiple ways the school librarian can assist in disaster preparedness or recovery through the use of technology. These technology-contributing roles are limited only by the librarian's technology expertise.

4. Community Organizer

Whether the librarian serves students and staff in a large urban school district or is part of a small community, the school's surrounding community at large needs to be a part of the decision-making process when student safety is at stake.

Librarians may volunteer in the role of community organizer related to the school's safety and emergency preparedness work. Each school's emergency preparedness committee likely has outside mental health officials and local emergency responders (fire and police). However, consider surveying your school and surrounding community to determine unique and individual-specific skills community members can offer. For example, is there a HAM radio operator nearby? Are there parents who have medical expertise or survival scouting skills who can contribute their unique talents or broaden the committee's perspectives for disaster response?

To protect students within the school, staff must also be aware of neighborhood risks, which may range from hazardous materials from nearby businesses or factories to local crime. In the role of community organizer, consider encouraging a community hazard hunt and assessment of neighborhood risks.

5. Reunification Procedure

One of the greatest problems that arises during any type of emergency is disorganized student pick-up, that is, parent-student reunification. During an emergency, no parents want to be separated from their children, and no child wants to be left behind. Without a preplanned and well-communicated reunification plan, chaos quickly ensues, adding greater disruption during attempts to solve immediate crises.

Chapter 8 on emergency planning offers a detailed explanation of how to establish a reunification plan and includes important documentation as well as reunification design. Librarians typically are familiar with every student, and students know and trust their librarian, someone who is less likely than classroom teachers to have sole responsibility for one classroom of students during a crisis. Taking on one of the many roles of leadership in the reunification process is a natural fit for librarians.

6. Supply and Resupply: Emergency Kits

All schools have emergency kits—one for each student—to help ensure student safety in a disaster (see Chapter 8). Schools design, assemble, and/or purchase these kits according to the culture and needs of the individual school and community, as well as geophysical concerns. While all students may be protected in this way, a school's day-to-day operations create constant changes such that when an emergency happens, these kits are rarely up to date. The librarian may provide welcome additional resources for whole-class emergency kits or may supplement kits for new students or when classes change enrollment. Librarians may also volunteer to conduct routine inventories of classroom emergency kits. Inventories and accuracy checks should be conducted at least three times during the school year: at the beginning of the year as kits are prepared or purchased, unexpectedly and unannounced midyear to remind teachers of their importance and ensure they are up to date both with content and matching student-to-kit ratio, and near the end of the year.

In addition, by the nature of the supplies used in the library, librarians may offer to provide additional supplies in the event of an emergency, particularly when students are forced to shelter in place for longer periods of time. For example, a librarian may provide a list of grade-level–appropriate class read-alouds for individual teachers (see Chapter 10) or suggest additional supplies, some of which may be found in the library supply source, kept in the library but available to teachers as needed.

7. Equipment Management and Inventory

Years ago, schools had a single student or a small group of students who served as audiovisual volunteers. These AV-expert students had multiple roles related to the management of audiovisual equipment. They delivered equipment, set up stations for viewing or listening, and helped manage the school's media.

Today, because of the omnipresence of technology, rarely does one student or even a group of students manage, maintain, and inventory all necessary equipment in the school. The management of technology requires many hands and the skills of technology professionals as well as assistance from the student body. However, it is not uncommon for the school library to be central in tracking some, if not all, technology equipment inventory.

Certain technologies are emergency preparedness specific, providing increased safety measures within the school. Further, equipment often requires peripherals such as cords, batteries, and chargers. In an emergency, these peripherals are critical for communication, security, and sustainability. Volunteering to house, inventory, and maintain these important safety technologies can provide a vital leadership role for the librarian.

8. Drills: Scheduling, Conducting, Evaluating, and Revising

Certainly all schools conduct regularly scheduled fire drills, lockdown drills, and other disaster exercises. Often, schools plan the drills for which everyone must evacuate the building on beautiful sunny days. Rarely are these drills administered during the most adverse weather conditions, when, in fact, an evacuation is more likely. Overall, drills are scheduled and announced well in advance and consequently do not replicate that which is more likely to happen: the unexpected "anytime" event.

As part of an emergency preparedness team, the librarian should suggest unannounced drills at various times of the day and of the school year. In addition, decision makers should be encouraged to assign specific roles and detailed scenarios for these drills. Ringing a fire drill bell and marching outside in line is no longer considered an adequate emergency drill procedure. Chapter 7 provides additional information on how the school community can more authentically prepare for disasters and other incidences that threaten school safety.

Finally, the school community must conduct a critical and detailed evaluation of the effectiveness of these drills and make necessary improvements based on its results. Librarians are encouraged to take the lead in implementing best practices for drills and their evaluation and revision.

9. Curriculum Delivery: Emergency Planning and Safety

This book's appendix on library curriculum offers sample lesson plans designed around various emergency preparedness and recovery situations. These lessons share instructional plans across grade levels and include a variety of instructional methods from highly engaging scenarios and role-playing to more traditional approaches to content knowledge. These instructional plans address a sample of emergency and safety topics as well as lessons to use *following* a disaster.

Each lesson comes ready to use with roles defined. Little preparation time is needed by the instructor (librarian) and when appropriate, lessons include assessment strategies and alignment with standards. Most include Common Core State Standards.

10. The Essence of Critical Time Management:

How does anyone in a crisis offer *time* as the critical contribution to safety and security? If the school has not experienced a major disaster before, most local members of the school community's emergency preparedness and safety committee may not be well versed in the importance of time related to response. Volunteer roles may come up in a moment's notice when time is most critical.

Offer to assist with recording and implementing time-saving elements for emergency planning. These may include writing up and updating email and phone trees or organizational command structure charts.

Consider simply offering to expedite information as needed when a crisis occurs.

11. The Information Specialist

Finally and most unique to this position only, the librarian holds the role of information specialist. Information distribution before, during, and after a disaster is critical. Items related to distribution of information include the ways information is gathered; the evaluation of key information before, during, and after a disaster; its range of importance involving actionable items; and above all, what to do with critical information, when and where. In preparation for and during a disaster, communicating information rapidly and efficiently can save lives.

LIBRARY LEADERSHIP

In his eloquent summary of the shifting vision of leadership for teacher-librarians in *The Many Faces of School Library Leadership*, David Loertscher presents a contradiction in that many individuals believe librarians are in "a relatively powerless position; yet, the same view is contradicted by the idea that information is power, knowledge is power, and wisdom is extremely powerful" (2010). At no time is this belief that information and knowledge are power more relevant than when using information and knowledge to prepare for their safety and keep them safe.

Loertscher goes on to suggest that previous stereotypes of librarians and libraries as simply a "container of information" have become obsolete. He further suggests the leadership librarians possess has always come from the "transformation of things and resources into high quality teaching and learning.... We transfer our leadership power to guidance and inspiration, and make ourselves indispensable" (2010).

In fact, Loertscher uses a natural disaster metaphor to illustrate the importance of leadership for librarians today. He writes that though some librarians have made the leadership transition, "Still others try to ignore the fact that their house is built on the beach and the hurricane is moving at a steady pace towards them" (2010).

During a crisis or disaster, it is leadership in the form of direction, guidance, and inspiration that will carry a school forward. And this leadership must be housed within a sound emergency communication plan, a strong emergency preparedness design, and a clear understanding by everyone in the school community of the importance of establishing a community grounded in respect for diversity, open-mindedness to differences in beliefs, and a strong sense of community—everyone should know and believe "We are all in this together."

Loertscher urges librarians to sit "at the table of the central governance of the school." And as librarians are asked to select roles of leadership within that central governance, assuming leadership for the safety and security of students and staff may well be the most important leadership role the librarian can choose.

TAKING THE LEAD WITH EMERGENCY PROFESSIONAL DEVELOPMENT TRAINING

Often, the professional development opportunities a school provides are scheduled a year in advance, sometimes even longer. These professional development release days often have a very "top-down" rather than collaborative decision-making process for planning and preparation. The agenda for professional development training for Columbine High School's school district changed dramatically after the shooting in 1998. It takes one strong voice within the school community to promote the importance of emergency preparedness and school violence prevention training for teachers *prior* to an incident, when it is needed most.

Certainly assessment practices, skills in rubric writing, and response to the Common Core State Standards are all important topics for professional development training, and it is not suggested here that these issues that currently have the nation's attention be trumped in favor of emergency preparedness and safe schools professional development. But there are methods by which *ongoing* professional development training, a more comprehensive and effective method for ensuring safety, can be integrated throughout the school year, and it takes leadership within the school community to make this happen.

Infused throughout this book are ideas regarding the impact that professional development training related to school crises and disaster preparedness can have. "Teacher-librarians can clear the pathway for teachers by providing staff development and procuring the tools, strategies, and access they need. They remove blockages by advocating for the needs of the whole school, not just specific teachers and students" (Gilmore-See, 2010). This is true in relation to both the school librarian's leadership role and the need for such leadership within the school community when and where it is needed most.

Audrey Church advocates the school librarian assuming roles of leadership for professional development in such areas as "in-service on students' use of web tools ... social media to connect to the local community.... [and reminding principals that] librarians are master teachers and collaborative instructional leaders in schools, with a broad, cross-grade knowledge of curriculum and of standards" (2013). With this diverse palette of skills, librarians are well positioned to integrate content regarding safety and emergency preparedness when delivering any professional development training sessions to the broader school community, achieving three significant goals simultaneously: demonstrating leadership, teaching integration of technology, and helping show how these tools can be used to keep students safe.

PERSONAL STORY: A Simple Disaster Exercise—The Pop Quiz

Some of the best ideas for preparing and sustaining preparedness in schools come from teachers themselves. Developing a disaster plan for a school is the easy part; training everyone on its contents is a greater challenge.

One teacher we were working with shared the following story: After the school's disaster plan had been developed, she was assigned to train people on its contents. After looking over feedback from the training, she noticed that most people said, "Looks fine to me!" From this, she—correctly!—assumed that most people had not read it thoroughly.

So she pulled what she thought were the 15 most important bits of information everyone should know. She turned these into 15 questions, made several copies of the plan, and when staff came into the room for the exercise, she passed out the "pop quiz" and provided copies of the plan, saying, "This is an open-book test. Work together to find the answers to the questions."

After 45 minutes, when everyone was done, they compared answers and discussed the plan. The end result was that everyone was now familiar with the plan's contents, knew the important points, and as a result of the discussion, realized some areas where they could make the plan better.

—Bill Lokey

TAKING THE LEAD WITH DYNAMIC EMERGENCY PREPARATION

While few schools today do not have emergency plans in place, some schools are so vigilant that they consistently repeat preparedness drills so that they will be well prepared should a disaster occur. Further, one of the greatest regrets communities that have suffered a disaster report is that they were not better prepared prior to the event. Following a disaster, those impacted always discover preparedness essentials that were overlooked, from small precautions to significant or broader community human and agency involvement—things they knew about, precautions they could have taken, but did not.

Librarians are uniquely positioned to help with crisis and emergency preparation for several reasons. First, because they are one of the few teachers within the school community with whom all students meet, librarians can provide every student with some baseline information, such as components of the safety curriculum and understanding of weather dynamics, thus providing consistency throughout the school. Additionally, librarians who serve as information specialists can provide leadership when information changes or when it needs to be updated rapidly, for example, movement of weather patterns when a natural disaster is imminent or changes in policy related to emergency response.

TAKING THE LEAD WITH DISASTER RECOVERY

Following a disaster, students and staff need to feel productive and will want or need to "do something" to create a sense of normalcy. Providing opportunities for research may help students and staff with recovery efforts, giving them the important post-event sense of purpose necessary for recovery.

Following a disaster, one of the most important things the librarian can provide is a place students feel secure, a safe haven. Part of this will be the facility (library) itself. Another means by which librarians can promote a safe haven for students is providing reading material—from read-alouds for teachers to individual student book recommendations—that help comfort and engender a sense of safety and security for all.

SUMMARY

Looking broadly at the various roles of leadership necessary in emergency preparedness and disaster recovery, key roles stand out as appropriate for the librarian. Deeper leadership may involve the librarian assuming chairing roles or subcommittee work. Particularly complementary matches include communication planning or hosting and promoting all-staff training days. The author encourages librarians to chair or assume a lead role in some components of disaster planning.

Here's why.

Disaster planning is similar to many other leadership roles librarians assume. The librarian is one of the few—if not the only—teaching members of the staff who meets with every teacher and every student in the school. In so doing, the librarian is a natural for the distribution of information and other key roles in disaster preparedness.

Librarians are well versed in sophisticated communication skills. During a disaster, the rapid, accurate distribution of information is critical; every member of the school community must be aware of whom to contact at a moment's notice.

When information needs to be prioritized or synthesized, the librarian has the skills necessary to manage this information.

A final word: Don't neglect to ensure home preparedness while preparing for emergencies at school. When taking on a role in preparedness, the librarian must model what is delivered.

Throughout subsequent chapters, some leadership role or roles will ring true and will seem a perfect match. The author encourages librarians to find leadership roles within the school's emergency preparedness and disaster recovery process that are that best match and to then work vigilantly to ensure safekeeping through that role. While helping to ensure the safety and security of all students, this role may become the way in which the school librarian is seen as the most important person in the school.

Stephanie's Story: Beach Elementary and Hurricane Katrina

To say that Stephanie Ferrell models the idea of Southern hospitality is an understatement. When I called her after she graciously offered to share her Hurricane Katrina story, it was in the middle of an ice storm. On the day we spoke, Stephanie was trying to maintain stability in her home, help her husband with issues around the ice storm's impact on his business, respond to the needs of her daughter with regard to the ice storm, all while ensuring she obey the town's restricted ice storm curfew. "Today the weather might get above freezing. If that's the case, water on both the bridges will ice over and cause more treacherous driving as the temperatures drop toward evening," Stephanie shared.

But the larger Hurricane Katrina story quickly took over our conversation.

Stephanie Ferrell was the librarian at Beach Elementary School, in Pascagoula, Mississippi, when Hurricane Katrina devastated the Gulf Coast in 2005. Beach Elementary is a small K-5 school located one block from the Gulf. To understand the proximity, Stephanie stated, "If you stand on the playground and look toward the beach, you can see the Gulf."

"Most houses in the area, including beachfront homes, were dramatically damaged or destroyed. My home was totally destroyed. The whole community was destroyed. We stayed at my grandmother's home for a while; only a handful of people were able to rebuild their home.

"Later, when people came back to salvage what they could, they found that everything was destroyed by mold and mildew," Stephanie explained.

"Most of the attention for Hurricane Katrina was focused on the city of New Orleans, which had the primary flood damage due to failure of the levees. In our community 92 percent of the town received some storm surge* damage. There were lots of feelings of abandonment from us with all the media focus on New Orleans. . . . from all of us, the rest of the Coast, that suffered extreme damage from the hurricane.

"For weeks we didn't have the simplest, basic necessities. We were literally homeless for weeks. You go from being a happy middle-class family to homeless. You have a mortgage, you still have to pay for your home, but there's no home."

The district encompasses two small towns and the destruction from Hurricane Katrina included 12 of the 13 elementary and two high schools. When they didn't return to school, those of the Beach Elementary students and staff who remained attended school at Central Elementary. Central had far less damage, although every school in the district except Lake School sustained damage. Central had only a foot of water from the hurricane. But even so, when classes resumed right after the storm, there were no doors, there were very few resources, and most equipment was damaged.

"We had to hang curtains over the bathroom doors for privacy," Stephanie explained. Any wood components of the structures that were remaining after the storm were quickly destroyed by mold and mildew.

"We were all so traumatized. So much chaos. For the first six to eight months, you're simply in survival mode," Stephanie explained. "It was hard to be there for the kids as we were going through the same issues in our personal lives and families. We were trying to take care of our students while trying to rebuild our own lives." FEMA made counselors available for both school employees and students.

"When I go down to the beach now and look at where I lived, it's very sad because there's nothing there."

Much of what proceeds following such a devastating event involves red tape, paperwork, and bureaucracy. Stephanie found that those in charge weren't necessarily prioritizing by who had the greatest need. No system was in place to ensure those with greatest need received assistance first.

"It was such an overwhelming disaster for the whole community that prioritizing was not an option. We were in survival mode. Everyone just showed up and did what they could. There were no definitive guidelines for rebuilding . . . most, if they had insurance, were denied the claim. Eventually federal and state agencies provided grants. Most insurance claims were considered 'damage from the storm surge,' so their claims were denied. Though it seems backward, when insurance companies start to distribute funds, the mortgage company gets the money first, so the people who had to abandon their homes received funds for rebuilding last. Some people went in and rebuilt, but only a few. Most of us lived in a small travel trailer for as long as 27 months. I knew people who had two or three children who lived in these small trailers. There were no places to rent, but these were people who had no alternative place of residence."

---

*Surge damage: The damage that occurs during the hurricane that is strictly caused by surf.

Ultimately the district did receive assistance for loss of property. Large federal and state grant funds were distributed for rebuilding schools.

"One thing we learned was the importance of documentation. If Beach Elementary lost three computers, they must be replaced with three computers at Beach Elementary; item by item, replacement was done this way.

"To receive the funds we had to rebuild Beach Elementary even though the Beach Elementary community was nil. One of our school families lived right next to the school. Their home was totally destroyed. They were gone, the school bought up the property, and we never saw them again. Never. Most of the population had lost their homes. They were forced to leave and find other residence. For several years we had fewer students because so many people had to relocate."

Students and staff from the school were displaced for nearly two years. The town itself received so much damage that a large number of residents were forced to relocate. Those that remained and rebuilt their homes are now attending Beach Elementary once again.

Beach Elementary Library: Then and Now "This Is My Legacy"

Prior to Katrina, the Beach Elementary School library had approximately 10,000 items in the collection. There were three sitting computers. The school was 60 years old. At some point the school constructed additions that were built to different building codes and therefore had a higher floor elevation than the older part of the school.

"In the addition a new very nice library was constructed; large, considering the size of the school, with 18-foot ceilings and a story time area. When Katrina hit, Beach Elementary School, depending upon which building you were standing in, was six to nine feet underwater.

"We were not allowed to go into the building right after Hurricane Katrina. Mold and mildew quickly took over, so it was difficult to determine the damage and begin the process for documentation of loss."

Following Hurricane Katrina, Stephanie spent a great deal of time pursuing resources available and writing grants. She wrote grants like crazy. She received a generous grant from the Laura Bush Foundation and another from the Dollar General Foundation.

Today Stephanie is refreshingly optimistic regarding the status of her school and its library. "Everything now in the library is just right. When I ordered books from grant money I received following Katrina, I was able to rebuild the entire collection from the ground up, and selections were made to align with the curriculum. Right now I'm writing another grant for eBooks."

"Now we have a 30 computer lab in the library thanks to the Dollar General Grant. Every book in my library is less than eight years old."

Stephanie's Advice Regarding Libraries Destroyed by Natural Disaster

I asked Stephanie if in retrospect there was advice she could give to school librarians who might potentially be struck by a large natural disaster. "There needs to be more than just a single plan in place. Have a plan in place for being displaced, one for temporary closure, and an additional plan in case of displacement. There's no way to be prepared for such an event, but the best thing I can offer is to keep very good inventory and be aware of all the resources available.

"Be very careful with inventory; inventory is critical. Everything now is web based rather than site based, so that is fortunate. Your data, your information, is in the clouds. When Katrina hit, we had all our critical inventory information on site, and it was destroyed.

"I tried to make the best of the downtime—our out-of-school time. Try to take advantage of these down times for keeping up with how the transition will proceed." During the time that she was displaced, Stephanie wrote grants and reviewed material for replacement of her entire library.

It's hard to imagine the devastation Stephanie, her students, and her community endured during and after Hurricane Katrina. Her story here provides us with excellent recommendations for responding to disasters. But—and perhaps more important—her story provides a testament to the strength of character that is needed in times of such unthinkable destruction.

Chapter 2

Natural Disasters

Before determining methods for responding to natural disasters, a clear understanding of what constitutes a "disaster" is in order. Emergency managers must often distinguish between emergencies and disasters. An emergency can be viewed as an event where individuals respond quickly to protect life and property and is usually initiated through the 911 emergency system. A disaster, however, is viewed as that which goes beyond one organization's capability to respond: when disaster plans are activated and extraordinary measures are authorized in response, when special instructions and legal procedures, such as officials declaring an emergency, are put into action.

Although this book sometimes uses the terms "emergency" and "disaster" interchangeably, both FEMA and the Red Cross distinguish among minor emergencies, major emergencies, and major disasters.

A **minor emergency** disrupts but does not greatly impact the day-to-day operation of the school. Examples include a flu epidemic, a short-term power outage, and a severe thunderstorm. The duration of minor emergencies and recovery is typically short, and they do not cause permanent damage to structures or school routine.

A **major emergency** may include incidents, events, and natural phenomenon that cause serious impact to portions or all of the school population. Major emergencies may stall, alter, or completely halt the day-to-day operation of the school. They include such events as flooding of the school's basement (which may house the school library) or a day of a severe temperature heat wave or cold spell.

In this book the term **"disaster,"** when describing a natural disaster, includes catastrophic incidents that occur on a broad or mass scale and impact to varying degrees the entire population of the school and beyond. Most often, natural disasters result in altering, sometimes causing permanent and major changes to, the day-to-day operations of the school. Recovery time for a major disaster is far greater than for an emergency. In the event of a major natural disaster, outside agencies and the school's surrounding community are involved. Thus, with natural disasters, plans that have been established (vulnerability assessment, risk assessment, communication plans, and emergency preparedness plans) are put into action.

One of the most important roles the librarian can assume in preparing for natural disasters is to help all students understand the phenomenon of weather and geologic processes and how they cause natural disasters. Whether a librarian acts within the school library alone or on a school-wide or community-wide basis, educating students about weather conditions and geologic processes that are likely to occur in the geographical location in which they are located is the best way for students to begin to be prepared for emergencies. It is impossible for a single teacher, let alone a librarian who may serve hundreds of students, to respond appropriately and have students respond appropriately to such situations if they have not been properly prepared for natural disasters.

The following weather conditions and geologic events may produce geologic, atmospheric, or hydrologic events that result in a natural disaster: **avalanche, drought, dust storm, earthquake, flood, heat wave, hurricane, landslide, severe winter weather (snowstorm or blizzard), storm (lightning, thunder, wind, and rain), tornado, tsunami, urban fire and wildfire, and volcanic eruption**. For each of these potential disasters, a unique set of conditions exists; therefore, the planning and response for each must be considered individually.

Most school personnel are familiar with characteristics that define these potential natural disasters; however, within the school context each disaster occurrence comes with its own set of considerations and understandings, both for preparation as well as response and recovery.

AVALANCHE

Compared to most major disasters, avalanches do not rank high on the list of threats to schools, and they are considered far less deadly and damaging than hurricanes or tornadoes. Yet every year thousands of avalanches are reported in the United States alone, and avalanches kill an average of 150 people worldwide and 29 U.S. citizens, annually. The majority of those killed (90–95 percent) are recreationalist males ages 15 to 35; 35 percent are snowmobilers, and 23 percent are climbers (Tremper, 2014) buried in an average of only 4.5 feet of snow (Spencer & Ashley, 2010).

Although schools are rarely constructed close to avalanche zones, many school-age students enjoy winter sports where avalanches occur. Some who are daredevils or risk-takers enjoy winter sports that involve skiing, snowboarding, or hiking "off the grid" or "out of bounds" from recommended recreational or groomed trails. Rarely do kids realize how often avalanches are actually ***caused*** by skiers and snowboarders.

Students need to know about the risks they can cause to mountain emergency responders and sometimes even untrained community volunteers who are called in to rescue those caught, trapped, or buried by an avalanche. "Research shows a human starts a snow slide in 90 to 95% of avalanche accidents" (Lanier, 2014), and only 1 percent of avalanche fatalities occur within controlled designated winter sport or lift areas. Yet, when youth decide to take up winter sports, no training is required for participation, and often they do not understand the potential risk and have little knowledge of avalanche warnings.

Historically avalanches have claimed a greater number of human lives than one might expect, particularly avalanches caused by earthquakes. In 1970, in the Great Peruvian Earthquake, 20,000 lives were lost as people were buried in mud below Mount Huascaran from the avalanche that ensued. Closer to home and more recently, in January 2000 an avalanche struck in Cordova, Alaska, destroying three homes and taking the life of one resident. In 2012 an avalanche caused by their own footsteps killed nine hikers in the French Alps.

Avalanches are sometimes predicted and often, in maintained mountain or winter sport resorts, avalanche experts throw dynamite or discharge cannons to force avalanches in order to keep winter sport areas safer for the public and maintain some control over time, impact, and size of an avalanche.

Temperature, wind, and degree of precipitation are factors that predicate avalanches. Powder avalanches, the least dangerous, are made up of lighter powdery snow blown or slid down a steep mountain incline; they occur following a heavy snowstorm of high precipitation. A slab avalanche, a more dangerous type that can travel up to 200 miles per hour, is caused by a snowpack containing a weak point. Slab avalanche is the type often caused by a skier or snowboarder.

Nearly all U.S. avalanche fatalities occur in the Western United States, particularly in Colorado, Utah, Idaho, and Washington (Spencer & Ashley, 2010). In schools close to mountain sport areas where avalanches have historically occurred, as part of the geography or information skills curriculum, students should learn of the dangers and potential risk of avalanches.

PERSONAL STORY: Avalanche Warning

Although I am passionate about my love of winter sports, particularly skiing, I have never been a daring out-of-bounds skier. Thus, I have not spent much time thinking about the threat of avalanches when skiing. But as I write this book during my annual stay in Sun Valley, Idaho, there have been three avalanches this month alone, in March 2014. As I watch the forecast on the Weather Channel, I view stories of an avalanche's deadly impact: "Four killed in avalanche . . . Two Colorado skiers killed by an avalanche . . . skier dies in Montana avalanche . . . avalanche hits Washington State resort chairlift." Eight people have died in avalanches this month alone across the country.

Only one of these was on the designated ski terrain. But as anyone who has been impacted by disaster will note: It only takes one.

—Christie Kaaland

DROUGHT

The planet is made up of 70 percent water. The amount of water remains the same: No additional water has been created, and no water has left the planet. Few, if any, natural disasters do not involve water in some form. This basic understanding is important to remember when considering that water's impact—*particularly with regard to natural disasters* - concerns not *if* but *where* water moves and where it is located. That is, both lack of water and an overabundance of water can cause a natural disaster.

Unlike some natural disasters such as earthquakes, disasters related directly to water—too much water or not enough—usually happen gradually over time. Such is the case with drought. This long-lasting weather phenomenon can last years, even decades, and typically cuts a wide swath.

Drought is most common in the states of New York, California, Colorado, and Georgia, but it happens not just in the driest regions. In 2008, for example, the Western states of Washington, Oregon, California, and Utah had mild drought that impacted crops and caused water shortages. In the past 40 years the impact of drought globally has doubled. In 2014 drought in the Western states, especially California, reached record levels of low water in reservoirs, causing wildfire and crop damage.

Those impacted by drought often focus needs and potential response efforts primarily on lack of water. However, other far-reaching impacts can also occur. In some regions where, for example, agriculture has a major impact on the lives of students, drought can have a long-standing and far-reaching impact on their lives (see the following feature about how drought impacts students). Such strange crises as rattlesnakes entering suburban areas looking for small rodents who have moved into buildings looking for water provide evidence that drought's impact reaches far beyond the lack of water.

How Drought Impacts U.S. School Children

During the 2012 drought season in the Southwestern United States, this drought's secondary impact had far-reaching results:

- *Mandatory water restrictions decreased children's access to water for basic hygiene, water sports, and potentially even hydration. (Mader, 2012)*
- *The impact on 2012 crops strained parents' and older children's employment opportunities. (Mader, 2012)*
- *During this drought "16 million children face[d] inadequate access to food over the summer, as breakfast and lunch programs are not part of a child's daily meals source" (Mader, 2012).*
- *An increase in separation, divorce, child abuse, and foreclosures were all noted during the 2012 severe drought season. (Mader, 2012)*

The impact and causes of drought vary from continent to continent. A lack of snowfall can prevent snowmelt from providing necessary water in streams and rivers. Wind can also cause drought from damage done to ground and soil, and it may fuel or speed up fires. The most famous of these, called the dirty thirties or Dust Bowl, lasted nearly a decade in the 1930s in the Midwest. Dust storms called black blizzards rolled across the Great Plains of Kansas, Oklahoma, and Texas. These storms that were heaped upon the region, coupled with excessive heat and the nation's financial disaster, the Great Depression, brought death and despair, causing many to wonder if it meant the end of the planet.

While opinions vary regarding global warming's contribution to drought, there is evidence that at minimum, some of it is caused by human activity and is the cause of the recent increase in drought on our planet. Our planet's temperature has increased by more than a degree over the past 100 years and is expected to increase another degree twice as rapidly, over the next 50 years. Some climatologists believe that there is a natural 22-year drought cycle that occurs in portions of the United States. They claim that this drought cycle, rather than global warming, has caused an increase in drought since the beginning of the twenty-first century.

Most climate experts agree that in some regions irresponsible rapid development has caused major drought. The Dust Bowl was partly the result of irresponsible agricultural practices. Fortunately, in the United States responsible practices such as proper irrigation, shelterbelts, and catchments are now required.

In regions where drought is most common, a consistent reminder to students of the importance of hydration may be the single most valuable contribution librarians can contribute to students' health.

DUST STORM

Dust storms are strong violent winds that blow sand and loose soil from dry surfaces in arid and semi-arid, typically flat land. As the wind picks up loose particles of soil, these particles bounce and hit the ground, causing more particles to be loosened and fly about. Weather conditions play a large part in determining the size, duration, and intensity of dust storms.

Though U.S. citizens rarely consider dust storms when imagining large-scale natural disasters, dust storms are more prevalent and more destructive in this country than one might imagine. Although more typical in arid and semiarid regions, dust storms cause far-reaching health hazards in the United States and are a significant source of airborne particles, particularly in the deserts of the Southwest. Most often, dust storms occur in agricultural areas where the soil has been manipulated or tilled bare.

Dust storms are one of the most critical weather concerns in Arizona. In 2013 in the second of Arizona's two major dust storms, a 19-car pileup caused by the storm's zero visibility took the lives of three drivers (Ferguson & Beal, 2013). Although fatality numbers may be small, this points to an important disaster phenomenon: often residents, and even weather, event, and responder professionals, focus on the primary victimization and manifestation of a storm whereby victims can be blindsided by secondary disasters' impact. In regions where they are prevalent, dust storms have become a major traffic concern as "walls of dust more than a mile high blanket an area in a matter of seconds, sometimes reducing visibility to zero" (Myers, 2011, p. 2).

As with most major natural disasters today, dust storms have increased in frequency, degree, and devastation. "During the 1960s there were eight dust storms that caused some serious damage; 13 more in the 1970s; 14 in the 80s and more than 20 in the 1990s" (Kidsworld, 2014). Scientists with the U.S. Geological Survey predict an increase in dust storm activity in the near future drastic enough to cause reduced vegetation. "[Dust storms] are no longer considered an episodic menace but rather a fixture of the landscape, the calling card of an emerging climatological crisis" (Abramsky, 2013).

The greatest destruction from dust storms happens because they appear suddenly without warning and can so rapidly increase and spread over hundreds of miles, reaching heights of 10,000 feet. Students with asthma and other respiratory ailments are in the greatest danger during dust storms, and the health risks to these students do not necessarily end when a dust storm subsides. Following a dust storm, small particles stay in the air, sometimes for weeks afterward.

The 2014 drought in California has left the Folsom Reservoir at record low levels.

EARTHQUAKE

One of the most significant considerations concerning earthquakes' destruction as well as preparation and

Bricks on the street after the 2001 Nisqually Earthquake in Washington State – Picture by Bill Lokey.

The blast from the eruption of Mt. St. Helens in 1980 knocked down trees many miles from the volcano.

response is that earthquakes do not give warning prior to striking; earthquake prediction is still an inexact science. In the United States an earthquake can occur anywhere, although regions where earthquakes are more frequent and more likely include the West Coast states of Washington, Oregon, California, and Alaska.

Further, unlike other natural disasters for which there may be a predictable season, particular weather patterns, or other conditions, earthquakes simply strike instantaneously and unexpectedly. Earthquakes are a natural geologic process and occur all the time. The U.S. Geologic Survey and many universities use seismographs to constantly monitor them. Fortunately, most earthquakes are too weak for people to notice, and most that are felt do little damage. On rare occasions, though, damaging earthquakes do occur. What this means with regard to emergency preparedness is that everyone must be earthquake prepared at all times.

An earthquake's seismic hazards determine the likelihood of harm or loss, either natural or human-made. Seismic hazard sources may also occur as a combination of the two; for example, landslides often occur as a result of an earthquake. Although the natural phenomenon of an earthquake may cause a landslide, building on fault lines, overconstruction, and other human-made building or construction decisions may result in greater damage and loss from landslides.

Before planning for an earthquake emergency, librarians should understand the school's seismic risks. In any particular region there are three factors that determine an earthquakes' seismic risk. The first is location. Particular regions of the nation are far more likely to experience earthquakes than others. The second seismic risk involves population density. The third risk involves all elements of vulnerability. In a school the vulnerabilities range from property and a school's building construct (such as the school's age and whether the building is up to current building code) to frequency of drills and preparedness resulting in greater, more effective response.

For a few days or weeks after an earthquake has occurred, while children may be startled or frightened by sudden loud noises, if no injuries or severe damages occur, for school children, recovery is almost immediate and they are less likely to suffer lasting impact or posttraumatic stress disorder (PTSD) than they would be from other disasters. However, because earthquakes occur with virtually no warning, educators have no time to help students make psychological adjustments to facilitate coping with the earthquake's results. It is recommended that children maintain school attendance or return as soon as possible following a quake to provide the benefits of peer support, return to routine, access to mental health services, and return to the productivity of school as diversion or distraction.

In the United States most injuries from earthquakes are caused by nonstructural items falling or breaking in the earthquake, such as shelving, pictures, windows, ceiling tiles, light fixtures, and unsecured heavy items such as television sets rather than buildings collapsing. Securing nonstructural items is not expensive and can be done with straps and anchors purchased at hardware stores or online.

Seismic Saturday

While I was the emergency management director for Pierce County, Washington, we worked with our local school districts on preparedness. One school district had a very active group of parents that took an interest in helping the district prepare. They started a project that they called Seismic Saturday, where in cooperation with the individual schools, they would go to a school on Saturdays and do nonstructural hazard mitigation. They would anchor ceiling tiles, secure shelves, add earthquake straps to televisions, put earthquake film on windows, and other tasks to make the building safer for students in case of an earthquake.

—Bill Lokey

FLOOD

Of all natural disasters, floods possess the greatest variance in cause and degree. Floods are the most common, most expensive, and most frequent of all natural disasters. Floods, like fire, can be caused by human error or natural weather phenomenon such as a hurricane or heavy rains. Flood damage may also result from fire and drought. For example, sprinklers set to automatically turn on during a fire often cause a great deal more damage than a small fire.

Two great concerns regarding floods are their unpredictability and great divergence. Floods can come on suddenly and violently as in a tsunami or hurricane or very slowly as, for example, a result of melting snow or excessive rainfall causing rivers to rise and gradually overflow their banks.

The chance of a school experiencing some water or flood damage is much higher than all other disasters except fire. The greatest likelihood for flood occurrence is from internal water damage, often from sprinklers or extinguishers. Further, and positive to note, the likelihood of human injury or death from *internal* water damage is less, relative, of course, to the amount of water, the speed at which it flows, and the extensiveness of the disaster.

One of the most dangerous, the flash flood, usually occurs without warning and moves at tremendous speed, often leaving a destructive path in its wake that destroys bridges, streets, and buildings as well as trees and other realms of nature. Flash floods are sometimes the result of dam failure, from which water can be especially destructive.

The greatest damage caused by floods involves other weather disasters such as the devastation from Hurricane Katrina. New Orleans sits an average of six feet (more in places) below sea level between two large bodies of water, making it a likely target for flooding, particularly following a major hurricane.

In the most devastating flood in U.S. history—the Johnstown, Pennsylvania, flood of 1889—a deluge of rain preceded the flood; the greatest damage was caused when the South Fort Dam broke, sending a 40-foot wall of water into Johnstown that killed 2,200 people.

As populations have increased and weather patterns have become more severe, flooding too has increased in severity and frequency. More recently, the mid-Atlantic Storm of 2006; the Tennessee 1,000-Year Flood of 2010; and the 2011 North Dakota–Missouri River Flood have impacted thousands of American schoolchildren. FEMA's website provides flood information (www.ready.gov/floods), and the American Red Cross provides a flood safety checklist which, at minimum, all schools should access and share (http://www.redcross.org/images/MEDIA_CustomProductCatalog/m4340128_Flood.pdf).

As Stephanie's story shows, floods can be particularly stressful for students when they take a great length of time to recede, thus having continuous and long-term primary impact on children. "Flood waters sometimes take quite a while to recede and the extended agony of waiting to see what is left after your home has been under water for a month aggravates an exceedingly stressful emotional situation" (Feinberg, 1999).

Educating all students about flooding—types of flood, flood zones, the science of floods, flood preparation, evacuation considerations—may produce critical positive results when they are needed most.

House along Puyallup River flooded in 1977. Note the high water mark. Picture by Bill Lokey.

HEAT WAVE

Unique to heat waves is consideration for the degree of intensity. Unlike many disasters for which occurrences are an absolute—one certainly knows when an earthquake, tornado, or hurricane has hit and when it is over—a heat wave can sneak up on its victims and may have no end in sight. Younger children, who are in greater need of frequent hydration, are the most likely to suffer health consequences from a heat wave. Heat stroke effects the youngest and oldest in the population first. Whether by children's own accord (e.g., being extremely active mid-day in the heat) or adults' unintentional neglect, dehydration and even heatstroke can happen quickly with intense heat. Constant hydration and coverage from the sun can prevent severe health risks during a heat wave. In U.S. schools, heat waves occur either at the beginning or the end of the school year, most typically in June, August, or September.

Although a heat wave is unlikely to do damage to a school library facility, during a major drought, librarians should watch for physical signs of its impact on children. Especially young children visiting the library may quickly experience severe dehydration. (See also "Drought" above.)

HURRICANES

Hurricanes have proven to be the most devastating of all natural disasters in the United States in lives lost. Since the year 2000, hurricanes have cost over 2,400 U.S. lives and billions of dollars in damage. Over 1,800 of those fatalities were in Hurricane Katrina. In the United States, hurricanes primarily hit the East Coast and Gulf of Mexico states, predominantly Florida, Louisiana, and Texas. Superstorm Sandy, the recent destructive force that hit the Atlantic coast in 2012, stands as the second largest hurricane on record. Unlike other disasters, hurricanes have a definite season that may stretch from May to November; they strike most often during September.

Fortunately, particularly since the impact is often so severe, hurricanes are one storm for which there is usually fair warning. Because of the consistent and lengthy warning period, called a hurricane watch, and the more intense hurricane warning, depending upon the track and severity of the storm, it is unlikely that children will be in school when a hurricane hits. What is more likely is evacuation for imminent hurricanes.

Over the past decade in this country, hurricanes have increased in intensity and degree of damage. Hurricanes are rated on a scale that was created by Robert Saffir in 1970 that measures sustained wind speed. Saffir-Simpson Damage Scale ratings range from 1 to 5. Category 1 hurricanes are labeled "minimal"; Category 5 hurricanes are called "catastrophic."

Hurricane Katrina was rated Category 3 (extensive), while 1992's Hurricane Andrew was rated Category 5. Recognize the complexity of Hurricane Katrina's damage, however. Hurricane Katrina was not only a hurricane; the majority of the damage that befell New Orleans due to this storm was

Flooded New Orleans. Picture by Bill Lokey.

caused by collapse of the levees built to sustain the ocean hurricane force of a Category 3 hurricane. Historically, Hurricane Katrina ranks third in number of lives lost.

Major hurricanes do not discriminate in destructive force. Katrina destroyed 110 of New Orleans's 127 schools (Horne, 2011). Today, few schools located in hurricane zones lack extensive hurricane emergency preparedness training for students and staff alike. Further, an abundance of research conducted following Katrina provides rich data for lessons learned and future hurricane preparation.

Generally, as Jeff Frank concluded in his thesis on Katrina's impact on libraries, "After analyzing [data], there appear to be two factors that impact the effectiveness of disaster plans for Gulf Coast libraries: the library's location and the scope of the disaster" (Frank, 2011).

10 Little Known Hurricane Survival Facts

- *Storm surge is the known danger, but half of all hurricane deaths occur from water and inland flooding (including landslides, mudslides, and flash flooding).*
- *The average person can be swept away in six inches of moving water, small children in far less.*
- *Water weight: One cubic yard of water weighs nearly a ton (1,681 pounds).*
- *Hurricane storm surges extend hundreds of miles inland, and it's often this unexpected surge that causes damage and death.*
- *23% of U.S. hurricane deaths occur to people who drown in, or are attempting to abandon, their cars.*
- *78% of children killed by tropical cyclones drowned in freshwater floods.*
- *Hurricanes can unexpectedly change course for no apparent reason.*
- *Hurricane season in the Atlantic peaks on September 10.*
- *If caught in or living in a high-rise building, stay below the tenth floor but above ground level.*

LANDSLIDES AND DEBRIS FLOWS

Landslides and debris flows (sometimes called mudslides) occur when large quantities of earth, rock, and other debris move down a slope or break away from a cliff. They are caused by any number of weather phenomena or human land modifications. Heavy rainfall, rapid snowmelt, rapid freeze-thaw, and soil erosion have all caused slides. Landslides also occur as the result of another natural disaster such as a tsunami, volcano, or earthquake.

Human-caused landslides are usually more predictable. At primary risk for landslides are areas greatly altered by construction, particularly where the land is steep or otherwise sloped. Mining, quarrying, deforestation, excavation, and some ill-advised land-use patterns have also initiated landslides; even the vibration from heavy equipment operation can produce slides.

On Saturday, March 22, 2014, a massive landslide occurred in Oso near Darrington, Washington, an hour north of Seattle. It took the lives of 39 residents who were tragically unfortunate enough to reside in its path. The mile-wide landslide produced a flow of mud and debris that clogged the Stillaguamish River, threatening flooding both upstream and downstream. " 'We have this huge square-mile mudflow that's basically like quicksand,' Snohomish County Fire District 21 Chief Travis Hots said. On Monday, ground rescue crews had to pull back from the area due to concerns about additional slide activity, the county said" (Berman, 2014). As with many U.S. landslides, the Oso slide will become further complicated by victims' guilt as previous knowledge of warnings of this event's potential are revealed.

Studying slides in school is complex; the full scope of information librarians can provide to help students fully understand the phenomenon of landslides ranges from ecological, political, and geopolitical, to engineering, meteorological, and historical (Bartley and Armstrong, 2014).

SEVERE WINTER WEATHER (SNOWSTORMS AND BLIZZARDS)

Snowstorms and blizzards can produce a unique set of conditions for a school community.

Typically, a snowstorm that befalls a community with strong winds, severe temperatures, and blinding snow is referred to as a blizzard when winds reach 35 miles per hour, snow impedes visibility more than 500 feet, and temperatures fall below 20 degrees Fahrenheit. Severe snowstorms are very predictable; in the United States, blizzards occur most frequently in Alaska and along the northern U.S. border states that stretch from Washington to Maine as well as the northern and central Midwest states.

A blizzard occurs when fast-moving wind currents (the jet stream) collide with warmer moist air. As this cold jet stream pushes down to meet the warmer air mass pushing up, a resulting storm brings high precipitation in the form of snow and sometimes ice.

The two greatest immediate dangers during a blizzard are impeded visibility and the extreme temperature's impact on the human body that may cause frostbite or hypothermia. Schools usually have warnings regarding severe snowstorms that result in the cancellation of school. However, when a sudden snowstorm results in children being stranded at school, power outages causing plummeting building temperatures produce the greatest danger for students and staff. Further, a shelter-in-place blizzard may result in students being stranded in the school for much greater lengths of time than for other disasters, sometimes even overnight.

Buses are often inhibited by impeded visibility during the snowfall. These conditions of blinding snowfall and large snowdrifts curtail not just buses but all traffic, including parents coming to pick up their children. Large deposits of heavy snow in a short period of time can also cause partial or entire building/structural collapse.

The likelihood and frequency of cold weather can affect a snowstorm's impact on a particular community. For example, in northern Alaska and northern Maine, communities are well prepared for snowstorms. However, a snowstorm that hits a populous southern locale can bring a city to a halt. On January 28, 2014, some Atlanta, Georgia, residents spent as long as 23 hours frozen in traffic when only four inches of snow covered the city. Many children and their teachers spent the night in their schools as transportation, both school buses and parent pick up, was impossible primarily because of traffic snarl. So, although cold-weather regions traditionally better prepare students for severe cold, Atlanta's 2014 winter storm demonstrates the importance of all communities implementing more comprehensive preparedness planning.

The most famous winter snowstorm impacting U.S. schoolchildren took place in January 1888 when one of the most sudden blizzards, known today as the School Children's Blizzard, occurred in the Midwestern United States. Mild morning temperatures did not prevent children from attending school, but as the weather suddenly changed, children were trapped in one-room school houses, and many died in the storm. This storm impacted most of the country; in New York City, 50-foot-high snowdrifts paralyzed the city, and over 400 people died. Fortunately, today increased communication and sophisticated weather forecasting better prepare schools for winter storms.

STORMS: LIGHTNING, THUNDER, WIND, AND RAIN

Greater in their potential for unpredictability are various severe weather storms outside the realm of hurricanes and tornadoes. While much current weather forecasting has reached a relatively precise state of predictability, storms can include great variance in weather fronts and cause damage via wind, lightning, rain, or a combination thereof. Greater understanding of the far-reaching potential damage from severe storms is an important precursor to preparedness for their impact.

Every year lightning strikes schools, and more frighteningly, tracks and fields where children are playing and practicing sports. Lightning annually kills an average of 85 people in the United States and another 250+ are hospitalized. Unlike many weather fronts, lightning storms occur in every state in the union. All members of the school community need to understand the "personality" of a lightning storm to decrease the likelihood of students being injured.

Interestingly, according to NOAA (**National Oceanic and Atmospheric Administration**) more than 80% of lightning fatality victims are male, between the ages of 15 and 40, and most lightning victims were heading for a protected area but waited too long to act. Reminding students that there is no safe place out in the open in a thunderstorm may prevent them from being struck by lightning.

> Remind children that thunder is only a sound and does not cause damage. This is important to share, especially with younger children, as it is often the sound of thunder during a storm that causes the greatest fear.

TORNADO

Tornadoes, like hurricanes, are strong windstorms, and though they are usually of shorter duration than hurricanes, tornado damage can be just as devastating. Although the warning time for tornadoes is not as accurate as that for hurricanes, some characteristics of hurricanes and tornadoes are similar.

Sometimes called twisters because of how they twist and turn as they move, tornadoes form when warm moist air, or updraft, meets cold dry air moving from another direction that is then set spinning by strong winds. These whirlwind storms spiral around a still center or "eye." Tornadoes blow the strongest winds of all storms and have been recorded at over 200 miles per hour with gusts greater than 300 miles per hour.

In 1971 Tetsuya Fujiko developed the Fujiko Scale to measure length, width, and speed of a tornado's destructive force. In 2007 the United States adopted an updated, more accurate scale, the Enhanced Fujiko Scale (EF0-EF5), to rate a tornado's strength on a six-point scale according to size, wind speed and gusts, and damage.

In the United States, tornadoes, like hurricanes, have their favorite locales, typically Midwestern flat plains states called Tornado Alley, although on occasion smaller tornadoes have touched down in areas across the country, from coast to coast. Tornadoes have struck in all 50 states; each year 1,000 tornadoes touch down in the United States alone.

More often tornadoes themselves do not kill or injure people; rather, flying debris is what causes the greatest injury and damage. For this reason, some school districts in tornado prone regions have begun stocking helmets as standard equipment in emergency preparedness kits (Breslin, 2014).

Tornadoes typically occur between 3 p.m. and 9 p.m., but on February 5 and 6, 2008, an unusual tornado storm event occurred in the United States. Called a tornado outbreak and later named Super Tuesday, this storm was unusual for its time of year, time of day, duration, and number of tornadoes. Beginning on February 5, over 92 tornadoes struck in 10 states, lasting from late afternoon until the next morning.

More recently, the most devastating U.S. tornado in over 60 years destroyed five schools and damaged four others, taking 158 lives when it touched down in Joplin, Missouri, on May 22, 2011. This supercell thunderstorm, which tore off the roof and broke all windows of Joplin High School, fortunately occurred on a Sunday when school was not in session. Schools built since this storm "have been outfitted with safe rooms, where students and staff can take refuge in case of a tornado" (Riedel, 2014).

TSUNAMI

A tsunami is usually caused by an earthquake or volcano that occurs under the sea that becomes a series of giant waves that travel thousands of miles, often accumulating speed and degree of intensity before crashing unto the land. Characteristically, a tsunami may travel at speeds up to 500 miles per hour and, as it approaches land, it will slow down, causing it to increase in height before crashing into the shore.

Tsunamis are somewhat rare—an average of four occur each year worldwide—and more destructive tsunamis usually happen only every 10 years. Warning signs of a tsunami include a major earthquake in a coastal area or a rapid rise in coastal waters.

A tsunami wave can reach 30 feet in height, and although rare, can double or triple this average such as the December 26, 2004, tsunami centered in the Indian Ocean off the coast of northern Indonesia that reached 110 feet, tragically proving tsunamis' power beyond the shore, that entire nations can be devastated by a tsunami. The countries of Thailand and Indonesia were most greatly impacted. Although this tsunami occurred in the Indian Ocean, most tsunamis occur within the Rim of Fire, the rim of Pacific Ocean coastal countries. This 2004 tsunami took 250,000 lives partially because the Indian Ocean has no major tsunami warning system in place like the Pacific region has. "'Nearly 500 schools were affected,' reported Sri Lankan librarian Upali Amarasiri (2005), 'which resulted in the loss of approximately 1.2 million volumes of books'" (Frank, 2011).

Total destruction from Tohoku tsunami. Picture by Bill Lokey.

It is no coincidence that the term "tsunami" is a Japanese term that means "harbor wave." The term originated from Japanese fishermen, who, upon returning from sea, reported witnessing devastation in distant ports. In 2011 one of the most devastating tsunamis ever recorded struck the coast of the Tohoku region of Japan, claiming nearly 16,000 lives. Although the term "tsunami" was once thought to be synonymous with "tidal wave," the latter term is inaccurate, as tsunamis have nothing to do with the ocean's tides.

Only schools located near the sea—and in the United States that is primarily the larger Pacific rather than Atlantic Ocean—have potential for being impacted by tsunamis. Tsunami trackers are able to predict tsunamis far in advance, and most school districts have community tsunami warning plans in place.

Washington State boasts the first school district to approve construction of a tsunami refuge. "Residents of Westport, Grayland and other communities in the Ocosta School District approved a $13.8 million bond issue earlier this year to replace a flimsy elementary-school building with a complex that includes a gym strong enough to withstand tsunami surges, tall enough to stay dry and big enough to shelter more than 1,000 people on its roof" (Doughton, 2013). The foresight of Superintendent Paula Akerlund, who worked diligently to convince her community of the safety value for its children, explained thus, "We're probably less than a mile from the Pacific Ocean, and we have no hills to run up or other natural high ground. Our only alternative is to get as high as we can, as fast as we can" (Doughton, 2013).

The schools in this coastal community practice drills unique to their community, tsunami drills, during the statewide emergency preparedness drill day called the Great Washington Shake Out. Many other coastal communities in the United States have worked on emergency plans for tsunamis and have earned the title of Tsunami Ready under a program from NOAA.

URBAN FIRES AND WILDFIRES

The two types of fire disasters to consider are urban fires and wildfires. Fires in homes and buildings, urban fires, are hazards common to all communities. Urban communities adhere to fire codes, fire drill requirements, and requirements for inspections. Wildfires, including forest fires, occur in open spaces and burn millions of acres of forest land, grassland and undeveloped land near urban areas each year. The most common place each can lead to a disaster is the interface between the two.

As destructive forces, water and fire have multiple personalities. A unique understanding of fires that rage through forests, parks, and wooded areas is imperative for children as they grow to become responsible citizens, whether a wildfire is caused by lightning or by human error such as a match or cigarette carelessly tossed, an untended campfire, or forest and brush wildfires.

When children experience a disaster involving fire, even without major injury or damage, the potential for PTSD or other emotional or psychological impact is greater than with some weather crises. Fire is frightening to a child. Fire, like an earthquake, rarely gives warning. Further frightening is the fact that fires lack predictability. In addition, fire has a potential for unlimited proliferation that other disasters do not. This may induce considerable fear, particularly in younger children. Further, children likely have seen visual images of fire and may have been warned of fire's dangers from adults in authority. These warnings are significantly important as lifesaving and necessary for child safety, but be aware that, counter-intuitively and surprisingly, these experiences of visual imagery and adult warnings will likely exacerbate a child's dramatic response to fire.

The causes of fires near or to a school are of greater variety, likelihood, and source than most disaster; fires can be natural disasters as well as human made. Approximately half of all school fires are arson, that is, intentionally set. Damage ranges from mild destruction caused by arson fires in trash cans to catastrophic fatalities such as the historic Lady of the Angels School Fire in 1958 set by a disturbed student with a history of arson, which claimed the lives of 92 students and 3 nuns.

Arson's prevalence demonstrates the importance of ensuring that all students know the limitless potential of fire damage and are well trained in fire prevention. Similar to the training students undergo for bullying prevention, students must know the importance of the bystander role in preventing damage from fires in their school: Catching fires early and reporting them saves lives. Such student involvement in prevention can thwart greater fire devastation such as the December 11, 2013, arson fire set at Turtleback Elementary School at which conscientious Cub Scouts immediately reported a student's suspicious activities, thus preventing extensive damage (Tenorio, 2013).

Most (71 percent) fires related to school and educational structures happen to school buildings themselves; the National Fire Prevention Association reported that from 2007 to 2011 a total of "5,690 structure fires in educational properties. These fires caused an annual average of 85 civilian fire injuries and $92 million in direct property damage" (Campbell, 2014).

Schools are more likely to experience fires during holiday seasons and major events. Christmas decorations, art and holiday projects, and even Thanksgiving food festivities create additional fire hazards not found during the rest of the school year.

During school fire prevention lessons spend time talking with children about fire. Letting them discuss fire in depth, both prevention and cause, is imperative, particularly with primary grade students.

VOLCANIC ERUPTION

The skies of southwestern Washington State blackened threateningly when the volcano of Mount St. Helens erupted on May 18, 1980. The eruption lasted nine hours and destroyed every living thing within 70 square miles of the blast zone. For several days afterwards, ash "rained" down on 11 Western states, causing health problems in people as far away as Idaho and Montana. Using sensitive seismometers and other measuring devices, scientists were able to monitor the behavior of Mount St. Helens prior to its explosion and provide information that led to a restricted zone around the mountain.

Over centuries of recorded history, from the 79 BCE eruption of Mt. Vesuvius in Pompeii to the 1902 Mount Pelée eruption on the Caribbean Island of Martinique, the magnitude of fiery explosion that characterizes a volcanic eruption has inspired myths, art, and music. But those impacted by volcanoes know that the raging power presented by volcanic explosions is unpredictable and can result in any number of health concerns such as respiratory illness, infectious disease, burns, as well as motor vehicle crashes and injury from falling objects. Further, volcanoes can cause wildfires, mudslides, and floods.

The impact from a volcano is not always limited to the explosion itself. Hazards from volcanoes include rock avalanches, ashfall (tephra), lava flows, and mudflows (lahars). On occasion a volcano may spill over into the sea and trigger a tsunami. In addition to the tsunami, heavy rainfall may occur following a volcano in which ash and rain create a river of mud trailing down the volcano mountainside, burying everything below it.

HOW CAN THE SCHOOL LIBRARIAN HELP?

Each weather-related phenomenon comes with its own set of adversities, conditions, and complexities. The next chapters provide school librarians and other educators with information on how to decrease catastrophe and chaos caused by weather-related disasters as well human-made disasters.

How can the school librarian decrease the likelihood of any of these natural disasters from causing damage, injury, or death? One of the most significant contributions comes naturally to the librarian: serving as an information specialist with regard to the unique circumstances of the community and the likelihood of specific weather conditions occurring within the school's vicinity. Certain characteristics exist for each weather phenomenon. The school librarian has multiple opportunities for informing, updating, and increasing knowledge of weather and its patterns, predictability, and idiosyncrasies.

At any grade, but particularly in the elementary school, where librarians often provide planning time for classroom teachers and regularly teach all students, each day is an opportunity for preparation, information sharing, and, when the unfortunate happens and disaster strikes, help with recovery.

Laura Pearle and Hackley School Fire

The school library at Hackley K-12 School in Tarrytown, New York, suffered complete destruction from fire that was caused when a lightning bolt hit the school early on the morning of August 4, 2007. Reports from the weather bureau that day claimed 200 air-to-ground lightning bolts had struck the area. Fortunately, students were not yet back in school.

Originally, the 100-year-old building that housed the library had been separated from other structures, but construction over the years had joined various buildings throughout the school. The fire that destroyed both the building and the collection for the Hackley School's library had housed all the library material for the K-12 students.

"We were able to provide exact documentation of what was in the library collection when the insurance company required it," Laura Pearle, the head librarian at the time, reported in a personal interview with the author (personal communication, January 9, 2014). Pearle had meticulously completed the collection's inventory just weeks before the fire. Because of this, she knew exactly what was in the library collection. Fortunately, the server was not located in the library.

"Keeping good records is critical any time, but when the catastrophic fire destroys your school, you realize how significant this is," Laura reported. "When the insurance company asked, 'What did you have?' we were able to record exactly what was in the school's collection."

Another recommendation that Laura made was having a good relationship with both the business office and the various departments within the school such as the science and English departments, with whom she worked closely.

While the insurance details were being worked out, the business office approved expenditures to ensure the collection rebuild went smoothly. Because of the excellent relationship she had built with those in her business office, she was able to begin reordering immediately.

"I was not the only one impatient to start the rebuild; the entire community supported it," Pearle explained.

She turned to vendors from whom she had quite recently placed orders and said, "Resend everything."

"They will bend over backward to help," Laura said of her vendors.

On rebuilding her library, including the collection, Pearle stated, "It was a full-time job just rebuilding the collection and planning for new space. I spent 30 to 40 hours a week for about a year rebuilding the collection." This work, done in addition to her full-time library position, was necessary for rebuilding her school library.

When asked what advice she would offer librarians from her experience, Pearle said, "My recommendations are to communicate with the community often about how the project is progressing, to keep meticulous records 'in the cloud' so when disaster strikes you have access, and to remember that this will take a lot of time and effort, so reach out to others when they ask if they can help."

Chapter 3

Human-Caused Disasters, Intentional and Accidental

Parents are going to be forgiving if test scores go down or the football team doesn't win, but if something happens to their child at school, they're never going to forget that.

—Amy Klinger, Educators School Safety Network

Being prepared for human-made disasters begins by having a clear understanding of the broad parameters of what constitutes human disasters and a working knowledge of how those events can impact the school community. Human-caused disasters include those brought about through myriad of direct or indirect human actions and decisions.

INTENTIONAL HUMAN-CAUSED DISASTERS

We won't be able to stop every violent act, but if there is even one thing that we can do to prevent any of these events, we have a deep obligation, all of us, to try.

—President Obama, January 16, 2013, at the memorial for Sandy Hook Elementary shooting victims

There are issues to consider that are unique to the disasters caused by humankind, whether from a terrorist or terrorist group, an outside invader, or member(s) the school's own population:

- Unlike natural disasters, no one is exempt from human-made disasters.
- Manmade* disasters are more unpredictable, which leaves the school community more vulnerable than for most natural disasters.
- With human-made disasters traumatic violence is more likely, and therefore. . . .
- Symptoms of posttraumatic stress disorder (PTSD) in students are more prolific, more likely, and may be more profound or long-lasting than those caused by natural disasters.
- On a positive note, much more can be done to help prevent human-caused disasters, particularly acts of violence within the school community.

So, while many of the plans laid out in preparation for a weather-related or other natural disaster apply to acts of violence in the school, with human-made disasters educators must prepare for additional considerations that include causal societal complexities. Those responding to school crises learn lessons from each tragic experience so that, at the very least, something learned may bring about change that can saves lives in the future.

Defining Terms

The same definitions for terms used in Chapter 2 to describe catastrophic incidents of natural disasters apply also to human-caused disasters, both accidental and intended. That is, **major human-caused "emergencies"** impact the day-to-day operations of the school; **human-caused "disasters"** usually

*While human-caused disasters can and have been inflicted by both male and female perpetrators, well over 90 percent are male perpetrators, so for the sake of brevity, "manmade" will sometimes be used to describe human-induced disasters.

impact the entire school population and may create permanent change in the routine and/or culture of the school.

I. SCHOOL VIOLENCE

Since the 2012 tragedy at Sandy Hook Elementary, U.S. educational, mental health, and protective services at large have worked tirelessly to make sense out of the unexplainable. To date, the most relevant lesson learned from tragic acts of school violence is that no school is immune; therefore, *all* schools must be proactive with regard to preparedness and vigilance.

Who Is to Blame for All This Bloodshed?

Of little good is finger pointing and blame laying without moving to act on the cause, and the first step is talking about it. Daniel Barron (1999), in remarks drawn from a presentation delivered to school librarians, asked his audience to talk about violence. He began by pointing to indictments at the far-reaching domain of prospective blame:

> It is the parents for not providing an appropriate set of values in their child's upbringing. It is the movie and computer game producers for pandering to the violent appetites of our young people. It is the Internet for providing information on bomb building, access to hate groups, and pornography. It is the NRA for equating hunting rifles with semiautomatic assault weapons. It is our government for preaching peace while at the same time slaughtering more innocent victims in a month than have been killed in all our schools in this century. It is the media for sensationalistic reporting leading to copycat crimes. It is the athletes, cheerleaders, and other favored groups for shunning and belittling the less appreciated and unpopular kids in the schools. It is society, for schools are its mirror image. Now, we can relax because we can identify those at fault and cast the blame.

Barron's point is well taken. Prevention begins with open, honest dialogue concerning school violence across an entire school community and a deep and focused examination of the culture of the school. One contribution the school librarian can make is to encourage discussions concerning the culture of the school and the frequency and types of violence that occur within the school, keeping such discussions at the forefront of the school's disaster plans.

Personal Story: Let's Talk about It

Throughout my teenage years my mother battled cancer, beginning when I was 12 years old. Through my junior and senior high years, as my mother courageously and repeatedly faced her cancer, I heard vague references: "stomach problems . . . ulcers . . . female problems," but no one ever spoke the word "cancer" aloud. I was not told that my mother had cancer until I was 18, only a few weeks before she died. It was customary at the time never to say the "c" word, "cancer," aloud.

Years ago Americans were uncomfortable discussing sexual harassment in school. Hundreds of sexual harassment cases suffered by students, even staff, were later reported by individuals who had tolerated much, from sexual verbal references to unthinkable acts.

Fewer than 10 years ago our nation's armed forces had a "Don't ask, don't tell" policy with regard to gays in our military, forcing some military personnel defending our nation to hide their own identity as an act of self-preservation.

*We have evolved as a nation to recognize that these traditions of silence were hurtful, and in response initiatives of policy changes have occurred. Yet, today the topic of school violence remains a difficult topic to discuss, and it is rarely added to faculty meeting agendas. I encourage teachers and librarians in schools to talk about violence—put it on the agenda for staff meetings and address concerns within the school—*before *a crisis occurs.*

—Christie Kaaland

Social Complexities Regarding Human-Caused Disasters

The frequency of incidents of violence in schools is staggering. According to the National Center for Educational Statistics (2011), in 2010 more students experienced violent victimization in school than outside school; students between the ages of 12 and 18 experienced 359,000 acts of violence committed against them in U.S. schools. During the 2009–2010 school year 85 percent of public high schools reported incidents of violent crime, totaling 1.9 million crimes (Neiman and Hill, 2011).

While Barron (1999) eloquently presents a broad spectrum of blame and influences, the complexities around *solutions* related to school violence are just as extensive. The issues stem from cultural, familial, political, and economic differences and often delve into deeper psychological, environmental, and even religious roots. Adults, including well-meaning school personnel, may potentially contribute to these complexities. Educators passionate about issues concerning violence in the schools, particularly with regard to gun violence, must not lose sight of the most critical factor of this complex issue, that is, *what can be done to help ensure greater safety for our students?*

Scores of well-meaning educators and researchers, from teachers conducting action research to school counselors, psychologists, and psychiatrists, as well as those in the medical, law enforcement, and child protection services field have sought answers to the complex issues surrounding school violence.

Opinions vary. Many people blame the increase in school violence on the abundance and severity of violence depicted in the media such as movies, video games, and music. Several studies have found that students who are exposed to extreme violence in video games are more likely to demonstrate aggressive behavior. But the question of whether this aggressive behavior actually causes or leads to extreme acts of violence and terrorism remains unanswered.

Others blame adoration of the American gun culture for leading youth, particularly in urban areas of high poverty, into adopting an adoration of guns. But while gun-inflicted homicide among urban youth of poverty is much greater than the rate in the general population, correlational studies remain inconsistent in their findings.

Media Violence

In an average week a child will observe 74 acts of violence through some form of technology: on television from both news reporting and fictional shows, on the Internet, and through video games.

> In news reporting, journalists are motivated by expeditious reporting, and often initial reports are inaccurate. In addition, motivated by sales and ratings, the media often tends to exaggerate for effect. For these reasons, youth must not be lured into believing all news reports. Teachers and librarians can help ensure that students understand this motivation. Librarians can teach children critical viewing skills that promote a more investigative, analytical spirit rather than an emotion-driven outlook when viewing the events of the day.

Movies, Television, and Music Videos

Science fiction films are not about science. They are about disaster, which is one of the oldest subjects of art.

—Susan Sontag

Over the past 50 years many studies have been conducted examining the impact on children viewing violence in the media. In one extensive study of 300 participants viewing 10,000 hours of televised programing, three primary effects of viewing televised violence were noted in the concluding report: "learning aggressive attitudes and behaviors, desensitization to violence, and increased fear of becoming victimized by violence" (Federman, 1998).

Wilson found, "From the research that has been conducted in the past, we can roughly estimate that about 90% of movies include some depictions of violence, as do 68% of video games, 60% of TV shows, and 15% of music videos" as cited in "Media and Violence: An Analysis of Current Research" (2013). In fact, movies and television shows that target children actually contain more violence than shows for adult audiences (Boyse, 2010). Further, the violence in children's shows, intended for children under the age of 12, is just as likely to glamorize violence as adult viewings. The average 18-year-old today has viewed 16,000 murders and 200,000 acts of violence (Boyse, 2010).

No one would argue that the media is a highly influential source of viewed violence for our youth, and educators have a significant challenge in countering the effects of the violence-saturation media culture.

Violent Video Games

> *While State's Attorney Stephen Sedensky III found no clear motive, he found items that suggest [the Sandy Hook shooter] had a fascination with mass killings, including a computer game called School Shooting that was found in his home.*
>
> —Blad, 2013a

One of the most controversial potential contributors to school violence is the proliferation of, addiction to, and overindulgence in violent video games by some youth. Although this media is relatively new compared to other media, today it is a multibillion-dollar industry, and though the majority of gamers are still under the age of 35, trends indicate an increase in older players as well as female participants. With regard to the influence of violence from video games, one alarming study compared older to younger players and found that "In relation to favourite aspects of game play, the biggest difference between the groups was that significantly more adolescents than adults claimed their favourite aspect of playing was violence" (Griffiths, Davis, & Chappell, 2004).

Although the Entertainment Software Rating Board (ESRB), established by the video game industry in 1994, attaches ratings to games based on violence, language, and sexual content, and although some states have passed rating laws, federal legislators have been unsuccessful in passing legislation restricting the sale of video games with violent content; opponents have won this legislative battle based primarily on free speech rights.

Over 300 scientific studies have been conducted examining the correlation between participatory violent video games and school violence. The most recent meta-analysis of research about video game violence (Anderson et al., 2010) explored the effect on physical aggression. Conclusions drawn from the 12 longitudinal studies that met criteria for best practices found at least positive relationships between "amount of time spent playing violent video games and later violent behavior. These studies, some of which were conducted internationally, included a total sample of 4,526 participants" (Media and Violence, 2014).

Schools have experienced increased school violence while, due to societal pressures, the past 10 years have seen a decrease in the production of violent video games. As with other areas of research seeking answers to school violence, conclusive evidence remains elusive.

Video Games in the Library

Public libraries are more likely to be impacted regarding purchase decision making with regard to violent video games. In fact, public libraries in Paterson, New Jersey, and Elmhurst, Illinois, have reviewed and reconsidered purchases of their video games rated "M" for mature (Jones, 2013). While many school libraries today provide both educational and entertainment interactive video games for students, they likely have far greater restriction on such purchases.

> Although it is unlikely school libraries offer video games with strong violent content, librarians recognize that their students have ample opportunity outside of school to engage in such games. Therefore, providing students with an opportunity to debate and critically think about the impact these games have may provide an opportunity to reflect on students' own and their peers' responses. Might such an open conversation also help uncover other issues concerning safety, respect, and student voice?

The Bully Factor

A University of Michigan Health System survey reports, "In the U.S., an estimated 160,000 children miss school every day out of fear of attack or intimidation by other students, according to the National

Education Association. Since 1992, there have been 250 violent deaths in schools, and bullying has been a factor in many school shootings" (Soulliere, 2009).

No one can deny that children today are exposed to a far greater number of violent acts than in generations past. At the same time, a generation ago many schools overlooked or tolerated bullying while innocent victims were bullied unmercifully; 40 years ago schools in the United States did not have antibullying policies in place, and it was not unusual for an administrator to use the phrase, "Boys will be boys" as he dismissed those few bullying acts that were reported, purporting that victims of bullying may need to "toughen up to face the world." From tragedies like those at Columbine and Virginia Tech, today's educators recognize the significant impact bullying has. Further, most schools have a bullying prevention campaign in place and, at best, children are trained to recognize the broad range of actions and words that define acts of bullying, and students are expected to report them rather than look away.

Violence from Outside the School's Population

In the heartbreaking narrative of Sandy Hook's tragedy, so many recognized warning signs came into play. The shooter was an isolated antisocial young adult with a history of mental illness, and "he very likely had at the least four of seven symptoms of schizoid personality disorder: lack of close friends, chooses solitary activities, emotional detachment, and does not desire close relationships" (Raine, 2013). He was obsessed with Columbine-type shootings, an avid violent video gamer, a lone gunman with access to an arsenal of weapons, and living alone with a gun-advocate mother (she intended to buy him a gun for Christmas) who was in denial of her son's worsening conditions and growing symptoms, granting him complete isolation.

This event has prompted many schools to look at broader perspectives, outside threats, and deeper issues, opening the door to dialogue with regard to all aspects of school violence. The ensuing dialogue has placed a greater focus on examining the culture of the school for improving students' safety and security.

Legislative Action

Over a decade ago the World Health Organization presented data on the then current state of violence and health and pointed out that violence should be considered a public health concern (Krug et al., 2002). Tragically, in analyzing data concerning youth-perpetrated violence in schools, the United States takes the lead in this health crisis. Recognizing this fact, the nation is leading the way in prevention strategies, particularly with regard to active shooters and targeted violence within schools.

In the world of scientific research, there is quiet movement in response to the outcry for help. The latest development came when the National Institutes of Health issued a broad invitation for grant applications to study firearms violence (NIH, 2013).

> We can't accept events like this as routine. Are we really prepared to say that we're powerless in the face of such carnage, the politics are too hard? Are we prepared to say that such violence visited on our children year after year after year is somehow the price of our freedom? (President Barack Obama, December 16, 2012, responding to the tragedy of Sandy Hook Elementary)

Following this announcement, the president introduced 23 executive actions designed to reduce gun violence, including support for the Centers for Disease Control and Prevention to study causes of gun violence. He also urged Congress to promote federal legislation on more clearly distinguishing regulation among types of firearms' use and more stringent federal background checks prior to gun purchase. Divisive politics around amendment-based rights, particularly the right to bear arms, have thwarted federal legislation. In Connecticut, where the Sandy Hook tragedy took place, however, led by Governor Malloy, voters approved legislation that creates statewide school safety council and competitive grants for improving security systems as well as greater restrictions on some assault weapons. Many other states have passed similar legislation since the Sandy Hook shooting.

In contrast, by April 23, 2014, 10 states had loosened gun restrictions including Georgia where on April 22, 2014, the most aggressive gun legislation to date was passed. "Guns Everywhere" legislation, as opponents have labeled this law, will allows citizens to carry firearms into schools and churches.

Librarians and classroom teachers may frame constructive dialogue around the issue of school violence within the context of legislation, promoting the importance of citizen involvement for seeking solutions and positive change.

Lessons Learned from Sandy Hook Elementary

Michael Regan, the former director of pupil services in Newtown, Connecticut, advises educators, "Prevention never delivers a guarantee and often leads to a false sense of security" (2014). Regan advocates that districts should not simply develop a disaster response plan but rather look at safety tasks as "planning," using a dynamic ongoing approach. Further lessons from Sandy Hook Elementary include:

1. *Invest in the people side of school safety (Trump, 2013).*
2. *Lockdowns are still one of the most effective tools available to get students and staff out of harm's way. (Trump, 2013)*
3. *To work, lockdowns must happen quickly. (Dorn, 2013)*
4. *Diversify drills and make them reasonable but progressively challenging. (Trump, 2013)*
5. *Locking interior doors to prevent outside entry works.*
6. *All staff in the building must be trained, not just a few.*
7. *If we want to decrease school violence, mental health issues must be addressed with equal attention.*
8. *"Pay attention to your community" (Swift, 2013).*
9. *Don't get caught in the battle over issues: "We're all on the same side in wanting to protect our children" (Swift, 2013).*

Addressing School Violence Is Everyone's Responsibility

The world is a dangerous place to live; not because of the people who are evil, but because of the people who don't do anything about it.

—Albert Einstein

While school violence perpetrated from an outsider is difficult to predict or prevent, much can be done *within* the school community to address violent behaviors, particularly bullying against and by students, as an approach to decreasing overall school violence.

Teachers and librarians work diligently to prevent violence and bullying and diffuse those who perpetrate bullying. However, large all-school antibullying assemblies, antibullying campaigns, and punitive measures alone have not proven to be effective in all cases. In fact, in some cases they have been shown to exacerbate bullying that is happening in the school (Pepler, Craig, Ziegler, & Charach, 1994). Further, "Youth who bully their peers are skilled at discovering and targeting the vulnerabilities of their victims" (Good, McIntosh, & Gietz, 2011), and classroom educators cannot be everywhere to prevent incidences.

Instead, some experts today suggest looking beyond the *acts* of bullying to a deeper understanding of the *reasons* bullying occurs. Some models for violence prevention that successfully decrease bullying integrate bullying prevention across the curriculum, and students are taught strategies to use when they witness bullying (Good et al., 2011). Good and colleagues found positive results both in the number of reported incidences of bullying and cases of students referred for bullying behavior, demonstrating the importance of students being part of the solution. This study was among many that point to the impact of empowering students to initiate cultural change within the school. An abundance of excellent antibullying teaching materials are available, and it is not the intent or purpose of this book to present solutions for bullying. Critical to solving in-school violence, however, is empowering all students to take an active part in preventing violence. It takes a village, and the school is that village.

Schools seek solutions to violence within the school from other influencing contributors by addressing substance abuse, sexual harassment, inequitable discipline procedures, and acts of prejudice and racism, all of which contribute to school violence.

Recognizing, Identifying, and Understanding Threats from Within

Active Shooter

Developing an Active Shooter section within the Emergency Manual should be a priority for your library and yourself.

—Kautzman, 2011

According to the Department of Homeland Security (DHS), with an active shooter, "The suspect's activity is immediately causing death or serious bodily injury; the activity is not contained; there is immediate risk of death or serious injury to potential victims" (2011).

As increased security measures work to prevent outsiders from entering school buildings, educators must also recognize threats within their own population, particularly threats from students.

Many resources are available to increase staff's ability to identify and understand potential student threats, and threat assessment has become one of the most effective methods of preventing school violence. Professional guidelines found in several studies conducted by various agencies provide a systematic, research-based process for assessing student behavior for potential risk of violence to self and others. Typically, multidisciplinary team members analyze a student's threat, including his past or changing behavior patterns, including current family status. Mental health professionals then make a "level of concern" decision, and intervention and support systems are put into place (for more information, see "Threat Assessment", Chapter 5).

Observing and reporting potential student threats is every educator's responsibility. Librarians often provide a respite for students who need escape from personal stress, school pressures, and sometimes persecution of their peers. In addition, the library is a facility open to all; it is not open only to students enrolled in a class. Further, libraries offer a central hub, a social gathering place, and a larger than classroom environment. These factors can create potential triggers for incidents such as the tragic Columbine school shooting, much of which occurred in the school library.

Understanding Active Shooters

While it is critical not to profile students according to personality criteria, identifying the following actions and understandings may contribute to educators' awareness with regard to active shooters:

- *Active shooters rarely, if ever, act without warning or prior planning; they do not suddenly "snap" but rather plan for months, sometimes years, prior. Conversely, nonviolent individuals do not suddenly change and resort to extreme acts of violence.*
- *Most active shooters have told someone of their plan and provide warning signs to those who interact with them regularly. Usually others in the school know about the attacker's plans, but the information is not reported to authorities.*
- *Active shooter incidents escalate rapidly and are unpredictable.*
- *Most active shooters have not previously directly threatened their victims.*
- *When active shooters have been stopped, usually someone at the school (rather than law-enforcement officers) stopped them.*
- *They are often motivated by revenge and not likely to respond to negotiating or reasoning.*
- *By the time they act active shooters assume they will die in the event.*
- *Following an event those who knew the active shooter well are rarely surprised by his action.*
- *Many active shooters have contemplated, written about, or attempted suicide prior to attacking.*
- *Often active shooters have suffered great difficulty with either personal loss or failure or else were bullied, otherwise persecuted, or injured by others in the school prior to committing the act of violence.*
- *Most attackers have not directly threatened their victims prior to an attack.*
- *Active shooters usually have access to weapons.*

Preventing School Violence: How Students and Peers Can Help

Schools that have established a culture of respect and safety encourage students to openly discuss their fears, concerns, and understandings with regard to school violence. Through discussion librarians and other educators can bring students to an understanding of their role and rights in preventing school violence.

> An adolescent comes to school with a collective life experience, both positive and negative, shaped by the environments of family, school, peers, community, and culture. Out of that collective experience come values, prejudices, biases, emotions, and the student's responses to training, stress, and authority. His or her behavior at school is affected by the entire range of experiences and influences. No one factor is completely without effect, which means that when a student has shown signs of potential violent behavior, schools have the capacity—and the responsibility—to keep that potential from turning real. (O-Toole, 2008b)

Prevention Measures

Soft Targets

On their website, the DHS states, "Most schools are still considered soft targets; this is due to easy access. There is little or no security at most schools" (Active Shooter Awareness Virtual Roundtable, 2011). More concerning is the fact that from 1966 to the present, nearly one-third of all active shooter incidents have happened in schools.

Schools should not be seen as a soft target by the DHS or any other agency or individual. There are actions and policies that can be put into place to decrease the likelihood of schools being easy targets. Assessing risks (see Chapter 5) and making changes based on the results of the resulting data will increase security measures. Seeking solutions to the school's physical, communication, technological, and people vulnerabilities (Chapter 6) is another. Each action taken to prevent violence is a step toward greater security and safety in our schools.

Actionable Security Measures for Improved Security

In fall 2013 a flurry of back-to-school media reports indicate that districts are making significant security system purchases—from adding high-tech "visitor management" systems that use Internet-based screening to checking identification cards for registered sex offenders and custody issues, to the latest video surveillance and communication systems.

Schools are fortifying entrances with bullet-resistant film on glass and adding panic buttons, door locks, and keyless entry systems. In addition, schools are budgeting money to hire personnel—from specially trained school police officers to security guards—to try to keep staff members and students safe (Molnar, 2013).

These measures alone will not guarantee safety. Security purchases must be partnered with addressing the school's culture to ensure the promotion of respect and inclusiveness, along with an awareness of students' mental health. "I had schools who used to have nine security folks there, and I put all that money into nine social workers. I saw a huge reduction in violence," said former U.S. Secretary of Education, Arne Duncan (Elliot, 2013)

Creating Codes as In-School Signals

Following a 2002 mass shooting at a school in Erfurt, Germany, schools began implementing code signals to alert teachers of potential violence. On March 11, 2009, in Winnenden, Germany, a recent graduate instigated a school shooting that ended in 16 deaths. Upon discovering the rampage, the school's headmaster broadcast the coded announcement, "Mrs. Koma is coming!" ("amok" backward). This pre-established coded message alerted teachers, who then locked their classroom doors. The shooter fled the building and continued shooting, but the headmaster's warning may have saved lives because it allowed precious seconds for police to arrive and interrupt the shooting.

As a preventative measure for outside intruders, establishing coded messages is a quick, effective communication tool (for more information, see the discussion in Chapter 4 about color-coding messages).

Security Measures

Early in 2013 school districts across the nation spent millions of dollars in attempts to increase security and protect children from harm. New high-tech security measures include:

- *Internet-based visitor screening*
- *Bulletproof student backpacks and other book bags*
- *Bulletproof whiteboards*
- *Internet-based security management systems for school websites*
- *Video surveillance*
- *More sophisticated communications systems*
- *Sex offender database identification*
- *Bulletproof glass and doors*
- *Panic buttons*
- *Keyless entry systems*
- *Entry-proof doorstops*
- *13-foot extension ladders (for escape)*

Producers of security-related items have also introduced a variety of briefcases marketed to teachers that provide the following options:

- *Pepper spray*
- *A hatchet*
- *Smoke grenades*
- *High-voltage stun gun*
- *First-aid kit*
- *Oxygen tank*

Positive Interactions

One of the ways schools can increase security is by developing a strong welcoming relationship with local emergency response agencies. Invite local police officers and firefighters to come to classrooms on a regular basis and to staff meetings. If such a relationship does not currently exist, start it in the library and suggest others in the school reach out as well.

Often responders are not used to being invited into the school except in response to a crime or other incident. Such an invitation may change the relationship between the school and emergency responders while providing greater safety visibility. Prepare them beforehand so that they understand they are being invited—welcomed—into the school and not to respond to an incident.

Although librarians may model this improved relationship by being the first to invite responders to visit, a one-time invitation is not enough to establish relationships and create ties to these community helpers. Additional suggestions include hosting a First Responder Appreciation Tea or establishing an ongoing Community Helpers Welcome Center where local first responders know they are always welcome. "Staff and patrons will know the names and faces of the police department and welcome the officers into the library. Their presence will dissuade troublemakers and allow for a healthier neighborhood environment where the police are seen as friends and safety officers, not harbingers of doom. If possible, locate a law enforcement way-station in the library so the police have a reason for stopping by. Their interaction with the locale will build a familiarity with the building, staff, and clientele" (Kautzman, 2011).

Librarians may wish to provide a student suggestion box—one that is well promoted so that students know about it—near the Community Helpers Welcome Center where students can anonymously post comments or questions for local first responders. Promote the suggestion box among the student population and offer anonymity as an option; students may have questions they would like to ask police officers or firefighters, or they may have comments they would like to make but don't feel comfortable sharing in a public forum.

For elementary schools, having first responders or law-enforcement officers read a story can send a powerful message of collaboration and security (see Chapter 10, "Books That Heal, Books That Help"). It may take just the right officer or firefighter to volunteer to read or tell a story to primary students, but once asked and they try it, these volunteers will likely return.

Also, that police car in the school's parking lot is likely to dissuade any outside intruder.

Countering the "Code of Silence"

In many schools there exists among students, particularly older students and even some adults, an unwritten but powerful code of silence. When threats of violence are made, many students feel that reporting these threats may implicate them. In addition, they may fear retribution. However, researchers have found that students attending schools where issues of violence have been addressed from an all-school perspective and that have worked to establish a culture of respect are more likely to break that code of silence.

Promoting a Culture of Respect to Replace the Culture of Violence

Schools are reflections of and directly subject to the conditions of the communities in which they exist.

—Barron, 1997

At the directive of the president, the Department of Education and the Secret Service joined forces in 2003 to examine evidence in school shootings to determine if patterns or predictors could be gleaned from these horrific acts to aid in prevention of future violence. The study found that no "profile" (i.e., a set of characteristics to look for) could be found across descriptions of perpetrators. Instead, the study determined that examining the culture of the school was more helpful in addressing both perpetrators and the school's potential for violence.

Schools that emphasize a culture of respect and personal connection among all students and between students and school officials have fewer violent incidents. In one California school at the beginning of the school year, students' names are posted, and students are asked to put a sticker next to the names of the students with whom they are friends. Teachers then work to connect those students with whom others have not shown connection, either by providing them with student connections or connecting with them personally.

Librarians are in a perfect position to practice this model because all students have access to the library. Helping make those connections can be done less obviously by suggesting that students post recommended book titles next to their friends' and classmates' names. This activity serves two purposes, as it promotes a culture of book sharing as well. The librarian may post suggestions next to students for whom no recommended book titles are posted so that no student is left out. Do, however, keep track of those students for whom no other student made recommendations and use this as a way to connect with students in need of a person who shows they care.

When there are indications that a student may pose a threat to the school either through information obtained via threats to students or staff, school administrators and law-enforcement officials must move quickly to investigate and intervene.

Communication Is Key

Throughout this text there has been a prominent theme of the importance of communication. While writing this book, I attended an emergency management and preparedness conference. The importance of communication was validated when the speaker on disaster preparedness for schools, in the introduction to his presentation, stated, "You're going to hear me say 'communication' 25 times in this speech. It's the key" (Esper, 2014). This holds true even more so with threats of school violence. Any threat of violence must be taken seriously and always investigated with the possibility of preventing potential attacks. Communication is the key.

—Christie Kaaland

II. ACCIDENTAL AND LARGE-SCALE HUMAN-CAUSED DISASTERS

Individual schools rarely establish emergency plans for large-scale human-caused disasters such as a chemical spill or a citywide pandemic. Instead, school districts—or more likely community-wide or statewide agencies—take over response and recovery operations when such events take place. But librarians who establish an awareness of potential local and regional hazards and an understanding of best practices when these major human-caused disasters occur can be invaluable assets to the school during critical moments. This section provides a broad but brief overview of other accidental and large-scale human-caused disasters.

Bioterrorism

Bioterrorism is the strategic act of using viruses, bacteria, or other harmful agents to intentionally inflict harm or death. Although the likelihood of large-scale bioterrorism in a school is less than most disaster threats, these threats should not be completely dismissed because the impact can potentially be so severe. Biological and chemical attacks can create mass disruption and panic. Bioweapons are easy to obtain and easy to conceal, and their contamination is difficult to confine. If a large-scale act of bioterrorism were perpetrated on a school in the United States, after the initial actions by local first responders, emergency operations would use federal law-enforcement agencies highly trained in international bioterrorism as responders.

Since the events September 11, 2001, the federal government has established several protective programs. Passed by Congress in 2004, Project: Bioshield Act provided funds to purchase and stockpile vaccines to protect against potential biological warfare. Funds were also legislated to establish several biodefense actions such as medical research focused on bioshielding against pandemic attacks and other preventative measures such as biosurveillance. Since that time, because of efforts in biosurveillance, the United States has been able to track potential terrorist threats and in response, countermeasures have narrowed acquisition of vaccines to primarily those that would be most likely needed—smallpox, anthrax, and botulism—with the number of doses increased annually.

Pandemic and Epidemic

The most likely epidemic a school district will experience is some form of influenza virus. Often the worst impact of a flu epidemic is disruption to the school year and high absenteeism.

A pandemic is an epidemic that has spread worldwide and gained global concern. Today with increased global mobility communities may experience a pandemic anywhere on the globe in short order. The most common pandemic is an influenza pandemic that occurs when a new strain of an influenza virus spreads. The danger of new virus pandemics is that everyone is at risk because no one has established prior immunity. Three major deadly pandemics have occurred over the past century. The

Spanish pandemic of 1918 killed between 40 and 50 million people worldwide. The Asian pandemic of 1957 killed 2 million; and the Hong Kong flu of 1968 took another million lives.

More recently as a result of the 2009 swine flu pandemic, between 8,800 and 18,000 deaths occurred. This is particularly concerning to educational communities because this pandemic disproportionately took the lives of children and young people under the age of 24. This is due, in part, to school-age children being in school with hundreds of others and therefore disproportionately susceptible.

Whether a disease causes an epidemic or spreads to become a greater pandemic, school closure and isolation (officially known as social distancing) are the most effective means of decreasing the spread of the illness. Recognizing this fact, school districts have established standards of infection for closing individual schools or an entire school district. For major epidemics or pandemics, mandates for school closure come from the state level. Whether directives to close schools come from state health departments (as 50 percent of U.S. states authorize) or state departments of education (37 percent), when such events occur, all health, education, and emergency management departments work collaboratively to inform citizens and respond to crises.

How Pandemics Kill

The number of people that die from a pandemic depends on several factors, including:

- *Number of people infected worldwide*
- *Severity of the strain*
- *How effective prevention measures are*
- *Amount/supply of antiviral drugs*
- *Vulnerability of the infected population*

It sounds simple, and it is simple: When an epidemic arises, before infection spreads or school closures occur, two easy yet effective steps educators can take with their students is to reduce the spread of germs through constant hand washing (for more information, see Chapter 8, "Emergency Planning, Emergency Kits") and contained coughing and sneezing.

PERSONAL STORY: An "Unabridged" Explosion

When I was in high school, everyone took one period of study hall that was held in the school library. It was in this class years ago—long before anyone identified such acts by high schoolers as acts of terrorism—that I encountered my first deliberate act of violence, one that in retrospect was quite mild.

Before the Internet provided details of such construction, a fellow classmate, through information garnered in his high school chemistry class, constructed an explosive device that he surreptitiously set up to detonate in the unabridged dictionary in that school library during my study hall period. The explosion was loud, unexpected, and frightening. No one was killed or even injured—except the dictionary, of course—but it's a vivid memory that has stood with me for years.

Today when I hear of disasters, both natural phenomenon and those caused by human folly or bedevilment, I am reminded of that memorable event. Yes, it reflects a sheltered childhood by current standards; it's a different world today. What I remember most is that it was loud, unexpected, and frightening.

—Christie Kaaland

Hazardous Material Contamination

A report by the National Center for Disaster Preparedness on the impact of the 2010 Gulf of Mexico oil spill disaster on children and families estimated that over 19 percent of the pediatric population in coastal Louisiana and Mississippi experienced emotional or behavioral distress related to the oil spill.

—Walker, 2012

Contamination from hazardous material ("hazmat") occurs in a broad range of forms, degrees, and proximities to schools, from simple cleaning chemicals and chemistry lab supplies to major hazmat spills. *Minor* exposure is one reason to keep hand sanitizer on desks or workstations to clean up and, after exposure to contaminants, wash hands. However, *major* hazardous material contamination near or at a school has a unique set of response needs.

Other than a few chemicals used in the chemistry lab, it is unlikely that an individual school will house lethal hazardous materials, and most school districts are required to include hazardous material inventories during safety inspections. These inspections include chemical inventory and identifying flammable chemicals that are then properly stored in appropriate areas in the building. Under federal hazard communication laws for the safety of workers and occupants, for any hazardous materials in the building, such as cleaning chemicals or chlorine for a school's swimming pool, Material Safety Data Sheets (MSDS) must also be available. These list hazardous properties of said chemicals as well as handling and safety information.

Should a hazmat contamination event *within* the building occur, students will likely be evacuated quickly, and medical responders will handle medical needs related to the contamination. In these circumstances, protective masks and gloves are needed. In the school library, having a fire extinguisher and cleansers for potential exposure may provide protection as necessary.

When a major hazardous material spill occurs *outside but near* the school, the type of spill determines response needs. For example, if the local water supply is contaminated, rapid and widespread warnings are key. Or if a vehicle carrying a hazardous material spills its contents near a school, the response would require rapid evacuation to a safer environment and medical treatment for students showing signs of reactions to the chemical.

Schools are unlikely to be located near an industrial plant that manufactures or uses toxic chemicals. Yet, if a major hazardous material spill occurs *outside but near* the school, such as the 2010 Gulf of Mexico oil spill, the far-reaching impacts may not be easily recognized. Because these materials are transported from industries via rail as well as trucks, their presence is more widely distributed than one might imagine. "The librarian role in the event of a hazardous materials accident might be to find information on the contaminant for the emergency officials" (Halstead and Jasper, 2005).

Under Title III of the Superfund Amendments and Reauthorization Act of 1986 (SARA Title III), also known as the Emergency Preparedness and Community Right to Know Act (EPCRA), facilities that produce, use, or store hazardous chemicals are required to report them to a local emergency planning committee. This public information, which should be part of the school's risk assessment, can be obtained through the local office of emergency management or fire department.

A simple lesson librarians can provide is to teach students, particularly younger children, to identify hazard pictograms. Recognition of such hazard stickers can provide a safer environment for students both inside and outside of school with regard to hazardous material.

War on U.S. Soil

Since September 11, 2001, tremors of fear have rippled across the world. [The World Trade Center] attack shook the foundation of our humanity. Now, the horrible images of war remind us nearly every day of the risks we face in a dangerous world. We are not the same, and neither are our children. As. . . . educators we face new challenges in our work because of the specter of fear that lingers at the borders of all our lives. Courage is persevering despite fear. It is gumption, grit, and the capacity to get up after a setback, with one's heart on fire.

—Smith, 2005

Although the 9/11 attack on U.S. soil was perpetrated by an extremist terrorist group and not a country, for the first time in over 60 years, U.S. citizens experienced an attack that represented the horrors of war. From this experience New Yorkers and others impacted by the attacks understood that while this nation is powerful, no nation is immune to attack.

Disaster preparedness for an attack of war on U.S. soil would elevate any emergency response implementation plans to the highest level of authority. Having emergency preparedness plans in place, including emergency supplies and effective communication, will provide the best foundational safeguard in the event of a national disaster such as war.

SUMMARY

The web of humanity spun its way in and around Newtown in the days, weeks, and even months to come—a silky web which at times brought such immense floods of love and support that it was overwhelming. People from across the globe reached out in prayer, love, and kindness.

—Connolly, 2013

Human-induced disasters weave a complex thread. Rarely do noticeable signs warn prior to human-caused disasters, whether intentional or accidental, which can leave victims more vulnerable than they are following other disasters and engulfed in trepidation. Educators, perplexed by the complexities of the disaster, are left stunned in the wonderment of how to cope and how to help students recover. Most move forward with the mission of helping children move into adulthood to become wholesome, sound, healthy citizens, and this must be the overarching goal for preparedness and recovery acts within the school.

Chapter 4

Communication

The sobering contrast between carefully written and rehearsed disaster operations plans and the actual performance of multi-organizational response systems under the stress of a catastrophic event like Hurricane Katrina demonstrates the limits of human capacity to absorb new information—especially dissonant information—as well as the need to review the technical and organizational requirements for supporting disaster operations in large-scale events.

—Comfort, 2006

Every professional disaster preparedness publication, presentation and website, repeatedly advises the public of the importance of establishing sound communication plans when planning for disaster. Further, following a major disaster, one of the common responses survivors report is the critical role of communication: both in having an emergency communication plan for the family and the workplace in place *prior* to an incident and adhering to those best practices in disaster communication *during* a disaster. Despite these repeated warnings and suggestions, emergency communication planning is often not given the attention and detail needed for effective implementation.

As a school leader who works with all students and staff, the school librarian may well be the most appropriate professional to take the lead in emergency communication planning. Within this subcategory of emergency preparedness planning are many roles of communication planning to which librarians can contribute.

CONSIDERATIONS FOR CREATING AN EMERGENCY COMMUNICATION PLAN

First, what is meant when we talk about a communication plan? Simply put, it is determining how to communicate with students and other stakeholders, including district administration, parents, the media, neighbors, other schools, suppliers, and local first responders. It also includes how **they** can communicate with **us**.

Designing and implementing a strong communication plan serves many purposes, including:

- Preparing the school community for timely and effective communication during crises
- Identifying potential gaps, misunderstandings, or weaknesses in communication strategies (see Chapter 6, "Vulnerability Assessment")
- Establishing inventories of and assigning roles to existing communication devices and technologies
- Managing distribution of information
- Empowering students and staff to respond to disasters and emergencies
- Meeting legal requirements according to state and local statutes

Communication is key to being prepared and establishing a plan, and rehearsing its implementation may provide precious seconds when they are needed most.

AUDIENCES OF COMMUNICATIONS

In terms of disaster preparedness, every school has four distinct audiences of communication, each with its own communication needs:

1. Students
2. Staff (all in-charge adults)
3. Families of students and the school community at large
4. Media and the general public

The safety of all students before, during, and after an incident requires a detailed emergency communication plan, with regularly scheduled training, drills for enactment, vigilant updates as needed, and improvements made based on best practices. If all of these considerations are addressed, staff will be able to respond quickly, to chains of communication command; families will know who to trust and where to go for information; and message systems to outside agencies will flow as smoothly as can be expected.

After a schoolwide emergency communication plan has been designed, the team must provide professional development within the school for implementing and sharing the plans. To present the emergency communication plan to school personnel, REMS (Readiness and Emergency Management for Schools) suggests conducting interactive role-play and re-enactment activities and providing example scenarios of potential disasters. Scenarios are available online from REMS (http://rems.ed.gov), or communication planning team members may choose to design scenarios unique to the needs of their individual school.

Incident Response Phases

In implementing the school's emergency communication plan during a disaster, there are often response phases that transpire. These phases vary depending on characteristics unique to the crisis or disaster. While some emergency response websites post general response phases, it is important to recognize how different these phases can be from one type of disaster to another. For example, Public Health Emergency (www.phe.gov) has determined these phases for communicating during a hurricane response: Phase 0 = steady state; Phase 1 = prepare (for normal operations, elevated threat, and credible threat); Phase 2 = incident response; Phase 3 = postresponse (recovery and mitigation). Hurricanes give lengthy warning and preparation time.

On the other hand, the unpredictability of a tsunami means response phases for tsunamis hold specific recommended actions at each phase, including longer watch time, "Phases 7: The threat may [or may not] continue for several hours after the arrival of the initial wave." (City of Santa Monica Tsunami Response Plan)

And of course some disasters, such as earthquakes and nearly all acts of school violence, provide no opportunity for anticipated actions.

Establishing an Emergency Communication Plan

Developing an emergency communication plan will save valuable time when time is of the essence. A sound emergency communication plan requires someone taking the lead in *developing* the plan, comprehensively *sharing* that plan through training and drills, and vigilantly *updating* the plan as changes in the school take place and best practices are modified.

The Emergency Communication Planning Team

As with all other parts of disaster preparedness, having representatives from various stakeholder groups will result in a more comprehensive plan. While it may not be possible to get representation from all stakeholder groups, at minimum, include a teacher, a parent, and, particularly in secondary schools, a student. Bringing in outside community members with first responder, medical, or mental health expertise adds value to message content. Consider also those with expertise specific to *communication skills* such as local media who are involved with community publications or those with writing skills. Finally, include community agencies with emergency response expertise and the local emergency management office.

Whether or not an administrator is on this team, once it has been designed, share the plan with administration. Regardless of who is on the planning committee, **there must be *no doubt* as to who is in charge when a crisis or disaster occurs**. Being a member of the communication planning committee does not imply being in charge during a crisis or disaster.

As one of only a few members of the school community with whom all students and faculty regularly interact, the librarian, who serves also as the school's information specialist, makes an excellent conduit for relaying emergency communication information. Thus, if the librarian is part of the initial emergency communication planning team, he or she can contribute to the plan from the larger school community perspective and will also have the deepest knowledge and a clear understanding of that communication plan when sharing with parents and outside agencies. Providing a link on the school library website for managing communication and delivering messages during a crisis or disaster may be one role to which the school librarian may commit.

I. COMMUNICATION ROLES

Before a clear emergency communication plan is designed, critical decisions regarding who is in charge during a disaster and clearly defined roles must be established, including but not limited to identification of:

- All stakeholders
- All communication roles
- A designated incident commander (IC), the "in charge" disaster authority person.
- A primary spokesperson (the IC may or may not also serve as the primary spokesperson.)
- The chain of command for notifying stakeholder groups
- Backup of incident command personnel (succession of authority) with delegated authority

Identifying all stakeholders who may or may not be part of the immediate school community may seem obvious. However, during the chaos of a disaster important personnel, sometimes even whole groups in line for communication, are often overlooked. Simply creating a chart of stakeholders and documenting the information in the communication plan may alleviate this oversight.

In **identifying communication roles** consider both those unique to the *individual school* as well as those unique to *individual potential crises or disasters*. Consider what expertise will be required of each role, as the expertise needed may help determine who will assume particular communication roles. For example, the emergency communication planning team may wish to assign those with technology expertise to roles related to implementing communication via technology such as social media tools, updating the school website disaster preparedness link, and videotaping or recording events. English teachers with writing skills may be called upon to script a template for the individual school's various potential emergency messages. The drama teacher may script scenarios for practice and drills.

The incident commander is the delegated authority who directs on-site operations until, and if, outside emergency response teams take over. **Identifying the designated incident commander** (IC) usually simply means ensuring everyone in the school community understands that the school principal makes all major decisions during a disaster. But this is not always the case. At times the principal may need to delegate his or her authority to someone qualified to take command in order to concentrate on the interface with outside agencies.

The administrator in charge when a disaster occurs may or may not elect to assume the role of the **primary spokesperson** both within the school and outside the school during and immediately after a disaster as well as in the days following, delivering messages to the community and media. The primary spokesperson must be an individual who is clearheaded in a crisis and able to calmly and succinctly provide information on what is going on and answer questions under pressure. The primary spokesperson serves to present opinions, give information, and clearly diffuse rumors, hyperboles, and speculation regarding the crisis or disaster. This spokesperson will make an official statement, coordinate messages with outside agencies, and respond to various interested parties. When designing the emergency communication plan, there should also be a designated backup person for this primary spokesperson's role. The librarian may wish to assume the role of the backup designee for the primary spokesperson.

Media Briefing for Primary Spokesperson

Whether serving as primary or backup designee, the spokesperson should consider the following suggestions when briefing the press or media:

- *Prepare a brief statement prior to addressing the media that covers the important facts of the incident: what, where, when, and who.*
- *Only answer questions that have been asked.*
- *Answer succinctly and clearly.*
- *You do not have to have all the answers.*
- *Distribute handouts, maps, and collateral material if possible.*
- *Keep cool and calm; appear comfortable, even if you have to fake it.*
- *Field questions distributively; one person may try to dominate.*
- *Do not speculate on "what if" questions. Stick to your message.*
- *Don't ever announce anything "off the record"; if it's presented to the media, it's fair game.*
- *Don't speak for anyone else unless they have asked you to and then use their exact words.*
- *Do not underestimate the power of the press.*
- *Do not underestimate the speed of social media.*

The emergency communication plan must provide the clearest pathways of communication. For most individual schools the **chain of command** begins with the principal, or, in the case of district-wide plans, the superintendent. The most important factor is to ensure that all members of the school community are informed in the most expeditious and comprehensive way possible.

Chances are in the case of a school emergency, the incident commander (IC) will be available and take charge. However, in the unfortunate event that the IC is unable to perform this duty, an alternative **backup commander** should be delegated with authority to step in. The assigned backup commander must be part of the initial emergency communication plan, not assigned during or following a disaster.

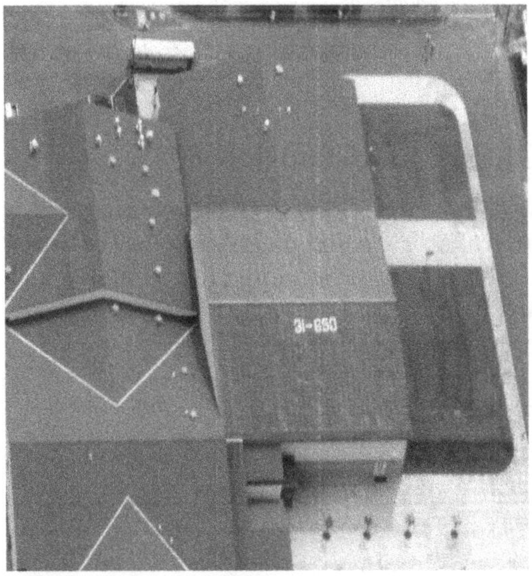

Numbers that identify the school and school district were painted on school roofs in Pierce Co., WA, to identify them from the air after a disaster.

Article from the Tacoma News Tribune about numbers being painted on the roofs of Pierce County WA schools to help pilots identify them after a disaster.

Further, all members of the school community must know who the backup commander is and when that person has taken charge of the school's emergency response.

Once designees of authority, communication roles, and backup personnel have been determined, the emergency communication plan team members must determine the means by which all forms of communication will be delivered. Here again, committee members should consider all possible scenarios, with backup plans in place. Any content the committee or spokesperson can compose as a template or describe prior to an incident will help expedite message delivery.

As horrific as the tragic December 14 event at Sandy Hook Elementary School was, the event could have been worse. Sandy Hook did everything right, and many school personnel performed acts of heroism. Sandy Hook Elementary faculty had gone through a lockdown drill just weeks prior to their tragedy. In fact, one of the classrooms still had paper from that drill covering the small window in the classroom door to prevent outsiders from looking in (Kamp, 2013). There was a closed-circuit television camera at the school's entrance and a buzzer for security. The school had a backup communication plan, and the front doors to the school were locked. All volunteers in the school were fingerprinted prior to being allowed to work with children. During the attack, the custodian ran through the halls and voluntarily communicated warnings, helping classroom teachers quickly lock their doors.

Tragically, the incident commander, Principal Dawn Hochsprung, and the school psychologist, who would later have provided professional assistance to grieving school members, were two of the first victims. This is a tragic reminder that it is never possible to plan for all inevitabilities when preparing for any crisis or disaster, but preparation can and does save lives.

PERSONAL STORY: Numbers on the Roof

During the time I was the director of emergency management for Pierce County, Washington, we had an active program of preparedness for our numerous school districts in the county. One concern was that if we had a major disaster, many of the aviation assets that would probably come to support us (helicopters and airplanes) would be flown by pilots and crews that were not from here and would not know what specific school they would be looking at if they flew over it. We found that some local pilots might be familiar with schools and hospitals in their neighborhoods but would not know them in other areas of the county.

Thus, we devised a numbering system for each school district and their respective school buildings. With help from school maintenance staff, local volunteers, and donations from a local paint company, distinct numbers were painted on most every school building in the county.

This way, if we had a major disaster and military pilots from other parts of the country flew over a school and saw people signaling, they could relay the number on the roof to the emergency operations center, and the location could be immediately identified to send help.

—Bill Lokey

II. MESSAGE DELIVERY SYSTEM: COMMUNICATION TOOLS

Though the human elements involved in emergency planning and recovery receive greater emphasis, also critical to a sound emergency response is analysis of communication tools and modalities. Part of establishing a communication system is determining what communication tools are readily available and how each will be used. Before a disaster occurs, conducting an inventory and analysis, and assigning roles to all available means and modes of communication, will improve response efficiency.

> "Ensure you have a way to communicate with the outside world. Charge your phone & have a solar powered cell phone battery." FEMA Alive blog.

Landlines

All schools, and usually all classrooms, have landline telephones with multiple lines available. Every classroom should have a one-page printed directory of all available lines. Librarians should include this directory of phone numbers in folders for substitute teachers and other personnel. If the chain of command has been established and well published, staff and students can identify where messages will be received via landlines and, additionally, how and what messages should be sent out from their classrooms. The Department of Education suggests having a pre-recorded or prescripted general "emergency" message for when phone lines become jammed.

Smartphones and Other Mobile Phones

In addition to landlines, most adults in the school and many students, even elementary school children, have cell phones. FEMA warns against the unnecessary and rapid depletion of cell phone battery life during a disaster, particularly when students and staff must shelter in place. This understanding should be an ongoing reminder and announcement on a regularly-scheduled emergency preparedness agenda item at faculty meetings. Ask teachers to remind students that should an incident occur, once they have texted family, their phones should be turned all the way off. If students and staff follow this suggestion, the distributive use of cell phones from the collective battery life can be managed from school's total cell phone arsenal for an extended period of time.

Depending on the incident and the expected duration of stay in a designated area, access to the Internet from one person's cell phone may be necessary. A primary rule of thumb is that within any group or classroom, replication of notification or access to information is unnecessary and can potentially deplete future needed cell phone batteries. Therefore, the person in charge—the librarian or classroom teacher—should, if necessary, designate one cell phone for Internet access.

> The teacher or librarian in charge must also caution students about taking photographs and, more important, posting or sending out photographs of incidents happening in the school. During a weather disaster, it is natural for students to want to take pictures of extreme weather conditions. For the most part, these recordings are harmless and may actually provide the school with recorded evidence for future use. However, in the case of an incident of school violence, remind students of the significance and potential impact of their recording and photographing thereof.

If students record or take pictures of incidents as they unfold in the school, hopefully, prior emergency preparedness training will provide an understanding of the value of using such recordings to help officials respond effectively. In addition, the recordings and photos could be used for future documentation. "It is good for law enforcement officials to be able to track the events leading up to and during the emergency through the use of [student] cell phone calls" (Anderson, 2008).

Student Use of Cell Phones during School Emergencies

For years most school districts, based on advice from safety departments and educational organizations, opposed any and all use of student cell phones for calling, texting, or taking pictures during the school day because of their disruption to the academic environment. Today, however, many schools, particularly secondary schools, have a BYOD (bring your own device) policy in place that allows portable devices for online access to research and access school academic sites during the school day. The policy should allow records of student and staff phone numbers so that the emergency communication plan can efficiently send emergency messages.

During a school emergency, student cell phone use has been found to hinder emergency response and potentially further complicate the situation. Hindrances include:

- *Hampering rumor control*
- *Inciting panic*
- *Promoting misinformation*
- *Overloading or increasing response team duties*

- *Impeding public safety response by accelerating parental response during evacuation*
- *Overloading cell phone systems during major disasters*
- *Hindering parent-student reunification*

Further, it is known that cell phones can also be used by students for calling in bomb threats; and, in fact, cell phones have even be used to detonate a bomb.

Despite this list of drawbacks, and while complete control of student use of these devices during an emergency is impossible, the most proactive strategy schools can implement is to educate students on the proper use of cell phones during a crisis or disaster. Creating and enforcing emergency guidelines can provide sound responses to student cell use during emergencies.

Two-Way Radios

Some schools currently use battery-operated walkie-talkies for communication across campus on a regular basis. When the school is without electricity, particularly for extended periods of time, such two-way radios can provide a valuable communication tool. Also, during major weather disasters some cell phone companies may not be operational. Thus, walkie-talkies supply multiple communication formats to help with recovery efforts.

When schools are in crisis communication mode, when there is not time to check batteries or especially to change batteries, having walkie-talkie batteries operational can be lifesaving. In some schools this equipment is saved for emergencies only. It is critical that the equipment be kept in flawless operating condition. One important role the librarian can contribute in an effort to ensure the school's two-way radios are always available and ready for use is making regular battery checks. Have plenty of backup batteries available as well in case of shelter-in-place or extended lock-down time.

Computers

Positive use and impact of computers and the Internet during a disaster are far more likely to outweigh the potential negative effects. However, teachers must caution students in their use of the Internet, particularly during incidents of lockdown when perpetrators are still at large or during lengthy shelter-in-place scenarios. If students have had curriculum content that has exposed the multiple capabilities of computers and the Internet related to efficiently accessing information and assistance during a crisis, as well as Internet safety curriculum, the Internet can provide valuable services and assistance in an emergency

III. COMMUNICATION MODALITIES

In addition to knowing about *types* of communication tools and technologies, one must understand the *modalities* by which communication takes place during a disaster or crisis and how messages and communication are delivered.

A. Computers and Other Technology

The most prolific communication modality for emergency communication used in the schools today is computers for Internet access for both individual and personal communication as well as delivering messages necessitating mass communication.

Computer Use: The School's Website

The Department of Education advises school districts to pre-emptively design an emergency website template on each school's website. Librarians should include a link on the library website as well. Announcing this link prior to disasters may provide an excellent far-reaching method for information delivery during an emergency or crisis situation, particularly with regard to the school's overall condition. For example, students and family members in a New Jersey school with an emergency website link that is suffering from a hurricane may not be equipped to inform grandparents in Arizona and California of their updated condition. With an emergency website template, family members across the country can at least view updates of their grandchildren's school, thus providing some relief for informing those at a distance. This can help alleviate jammed lines of communication such as student cell phones as well as provide relief from the stressful "unknown" during severe weather emergencies.

Computer Use: Social Media

There are no hard and fast rules for communication modalities related to various social media when applied to communication during a disaster. Children will tweet. No school can prevent every individual in the school from posting announcements during a crisis, particularly during a lockdown when students are confined. It is amazing how quickly today's youth can create a YouTube or other online multimedia view of the events currently taking place. Rather, school officials should address the use of social media tools for emergency situations from a perspective of decision making and problem solving: protecting the lives of all members of the school community.

> As part of the emergency preparedness team, librarians may volunteer to be the social media monitor or instructor. While it may be impossible to prevent postings, having someone watching what is being posted may inform teachers and administrators of that to which they may soon need to respond. Misinformation can be quickly discounted and rumors de-escalated. Overall, being proactive is critical; social media must not be ignored during a disaster.

The librarian is in an excellent position to deliver Internet safety curriculum content on the appropriate use of social media tools related to a disaster or crisis. This content can be integrated with other subject area content, taught in collaboration with other classroom teachers, or delivered as part of the school's disaster preparedness curriculum.

In addition to social media concerns around student use, each school should establish policies related to staff members sending incident announcements, both internal and external, during and following a disaster or crisis, via Twitter, Facebook, and other social media tools. As the incident commander provides official announcements, the school librarian as the social media designee may serve as an appropriate conduit for forwarding or delivering announcements and messages on social media. It is critically important that the messages from official sources are consistent.

Computer Use: WEAs

> Librarians may wish to volunteer for this one simple but very important role: enroll and deliver Wireless Emergency Alert (WEA) messages. Librarians can receive early warning signals from FEMA by signing up for their public service, Wireless Emergency Alerts (WEA), which is available via the FEMA website. "Wireless Emergency Alerts are emergency messages sent by authorized government alerting authorities through your mobile carrier" (FEMA) alerts are available free for all newer-model cell phones.

Important advantages of signing up for WEAs include:

1. The message, issued from the government, will have reliable and accurate information.
2. Cell phone users have continual access to these alerts.
3. Only significant and locally important alerts will be sent.
4. Warnings are no more than 90 characters.

5. Follow-up details will include instructions on actions to take, date and time of the alert, and the agency distributing the message.
6. WEA messages are not affected by network congestion.

The three types of alerts sent via WEAs are (1) those concerning extreme weather, (2) AMBER (America's Missing: Broadcast Emergency Response) alerts, and (3) national emergencies, which are usually announced by the president. Particularly important for educators, AMBER alerts provide any known details about abducted children. Weather alerts are particularly important in regions where action must be taken to survive severe weather such as tornados and hurricanes.

Computer Use: Personal Emails

Students are more likely to use their cell phones to text or Tweet than to send personal emails, particularly during a disaster. However, staff who understand the value of creating custom lists for informing groups via emails may create various listservs ahead of time that are then ready to deliver messages to each predetermined group. Accurately labeled groups can save valuable time when critical announcements must be made. Also, only those on a designated list will be notified. For example, emails to an Emergency Response Agencies listserv can deliver messages to all local agencies; an Emergency Response Team list can initiate a pre-established phone tree; and Parent Emergency Team Volunteers can pull in members from the outside community who have received appropriate prior instructions and who are ready to volunteer.

Computer Use: Texting

Texting, as the current student form of passing notes in class, is a ubiquitous activity that must be integrated into the emergency communication plan. Students *will* text. Establish standards, distribute appropriate messaging, create an emergency message FAQ template, and integrate emergency response texting policy with other school policy training. The more ways the school helps students understand the shared responsibility of emergency response, the more responsible their use of texting will be during times when cooperation is most critical.

B. Television and Radio

The school's primary spokesperson should communicate all external messages and news briefs during and immediately after an emergency. While it may not be possible following a disaster to alleviate students' anecdotal interviews and responses to high-profile incidents, if students have been trained in what is appropriate to share, they are more likely to respond in a sensitive and professional manner. Students should be requested to refer members of the press, including television and radio reporters, to the school's primary spokesperson.

Media Response Center

Prior to a major disaster or other major incident, it is suggested that the school designate a location and time for potential media response. This designation will help prevent unnecessary media hounding and may provide relief for stressed or grieving school personnel. The media response center should be coordinated by the external primary spokesperson. It may be appropriate to locate this response center some distance from where the incident occurred.

> When media response is needed, it may be a time the librarian can draw from library training and expertise to help craft incident response messages based on specific facts of the incident. Volunteering to help craft a spokesperson's message will come as a welcome relief. It is unlikely that a school spokesperson will have much prior experience addressing the media. It is critical, for legal and emotional reasons, that when dealing with the media, the message be clearly scripted in the form of a statement.

Be sure to set a time limit for media response. Leave no more than a few minutes for questions. The Department of Education's Office of Safe and Healthy Students suggests having a prepared statement recorded on a USB drive or CD and delivered thus.

C. Technology-Free Communication Modalities

Adult and/or Student Couriers

As the shooting continued, custodian Rick Thorne rushed through the halls issuing urgent warnings to anyone who had not realized what was happening. As he passed each classroom he locked any door left open. Instead of heading for safety, Thorne remained in the hallway and stood guard outside the locked classrooms as the shooter prowled the school.

—Owings, 2014

While the rarified heroism demonstrated by Sandy Hook custodian Rick Thorne certainly goes beyond expectations of staff communication, his story serves to demonstrate the unlimited potential of using any and all methods of communication during a crisis.

Many school emergency incidents that require instant communication are not conducive to allowing adult and/or student couriers to deliver messages throughout the school. When the school is under attack or there is an active shooter in the building, for example, courier delivery is likely too dangerous.

However, at other times the use of couriers can be the perfect choice and may have advantages over other communication formats:

- Younger children will not be alarmed by a known adult or older student entering the classroom to deliver a message. Students trained as couriers may be coached to appear calm, unruffled, even stoic when delivering messages such as potential school closures due to imminent weather fronts.
- Unlike use of some technological communication tools, couriers can deliver messages to exactly the people who need them. For example, if the school counselor is in a classroom but is needed immediately, a courier can deliver a note only to that counselor with specific details, and no one else may be the wiser.
- No electricity is needed, and no equipment failure can happen.

Handheld Bullhorns or Megaphones

Use megaphones for delivering quick, short (usually one-sentence) messages or information when no time for discussion or explanation is needed, for example, "All fifth graders proceed immediately to the cafeteria and await further instructions."

Such equipment can be useful when electricity is unavailable and mass and immediate announcements or messages need to be made. It is important to note that the use of bullhorns may alarm younger children and can be quite frightening; thus, careful consideration regarding tone and wording of the message is important.

Bells and Fire Alarms

Discussion of bells and fire alarms is included here because most alarm systems have backup capability when electricity is down. In some schools the alarm system has various forms of communication beyond simply a release from class. If the alarm system is a well-established form of communication, creating messaging using various tones to represent levels of alarm, for example, for a natural disaster, can provide an immediate and comprehensive warning system.

> "Text, don't talk. Unless there's a life-threatening situation, if you have a cell phone, send a text so that you don't tie up phone lines needed by emergency workers. Plus, texting may work even if cell service is down," FEMA advises (fema.org).

Emergency Backup: Shortwave, Amateur, and HAM Radios

Who hasn't heard fascinating stories of mystery, intrigue, rescue, and survival via communication due to amateur radio operators' bravery? Very little is needed to communicate via shortwave two-way radios. In a severe weather situation HAM radio operators may provide critical communication. In times of emergency it is valuable to know who local HAM operators are. Consult the Amateur Radio Relay League or National Association for Amateur Radio website (www.arrl.org) to locate nearby HAM operators.

A final note with regard to communication modalities: In the event of a major disaster or crisis, the more forms of communication by which emergency procedure messages can be distributed, the more likely procedures will be handled smoothly and comprehensively. With repeated drills using varied communication tools, procedures potentially become second nature to members of the school community. Also include these procedures in all standard school publications such as a students' rights and responsibility handbook, new school orientation publications, and back to school letters and emails.

IV. ALL CONSTITUENTS ON BOARD

While each constituent group message is worded differently, communication planners must remember to include everyone in the broader school community when considering communication before, during, and after a disaster. This may seem obvious, but when emotions are high and people are under a great deal of stress, even the most obvious plans and decision making can become difficult, which is another reason having a plan is so important.

Color-Coding Messages

To mask messages delivered over any all-school communication system such as the PA intercom system, many schools use color-coded messaging systems. Color-coding can prevent young and vulnerable children from hearing disturbing, upsetting, or confusing messages. It should be noted that although color-coded systems prevent outsiders from being privy to school policies and procedures, the more complex and detailed emergency communication procedures become, the more likely confusion will occur. Further, it is unlikely that visitors to the school such as substitute teachers will understand codes or know what to do when coded messages are announced. Schools that choose to use color-coded messages must include them in emergency drill procedures as well as all substitute folders and new employee guidelines. The following list provides an example of a color-coded messaging system used in school district communication:

Code RED: External violent threat, terrorism, school under attack
Code ORANGE: Internal active shooter, on site/on campus
Code YELLOW: Phone-in bomb threat
Code GRAY: Potential approaching severe weather front

V. MESSAGE DELIVERED LOUD AND CLEAR

Once the chain of command has been established and roles and modes of communication determined, the communication planning team should use graphic organizers to create a phone tree, emergency information sheet, and a flow chart describing how the flow of information will be delivered during a disaster or crisis. Follow up this planning with multiple comprehensive training opportunities. While a phone tree or message flow chart sounds basic, a single loophole in the flow of information or messages can have devastating results. Therefore, reenact the flow of information multiple times throughout the school year to help improve the likelihood of rapid, comprehensive, successful communication delivery when it is needed most. It takes only a few minutes at a staff meeting at which most adults in the school are present.

Emergency Communication Plan: Message Content

With some exceptions, for example, what to do if a tornado is imminent or if there is a fire in the school, most emergency message content cannot be preplanned or composed prior to a disaster or crisis;

every incident is unique, and when disaster strikes communication must be accurate and specific. However, to quickly prepare messages during an incident, communication planners should practice drills and scenarios with all relevant agencies, groups, and other stakeholders. Think through all the roles that need to be filled with regard to communication during an incident and plan drills accordingly. While it may not be appropriate for the exact message content to be scripted prior, communication planners can compose prescripted messages or templates for communication during a disaster for various targeted audiences and adjust for various types of disasters. The more opportunity for reenactment of potential incidents, the more likely things will run smoothly during an actual disaster.

Text 911

Probably one of the quickest and most student-friendly emergency response methods used today is texting 911. This provides broader emergency response modes (e.g., for individuals with a hearing or speech impairment), and it is valuable when it is dangerous to call on a phone. "Starting on May 15th the four major wireless carriers—AT&T, Verizon, Sprint and T-Mobile—are making it possible to send a text message to emergency services 911" (Das, 2014).

Emergency Contact Information Form

911 is not enough. Every adult in the school community should have a printed list of emergency contact information at the ready at all times. At the very least this should include local emergency response numbers such as fire department, police, and hospital. American Association of School Librarians (AASL) Beyond Words site has an excellent downloadable form that can be accessed at http://www.ala.org/aasl/awards/beyond-words/site for proactively recording these numbers; always follow up electronic versions with printed copies as well.

Determining the School's Efficiency for Emergency Procedure Communications

You can probably predict how effective communication within your building is. The librarian has access to all students and all staff, thus granting both greater responsibility and greater potential for ensuring effective communication. Read through the following list to determine your school 's communication effectiveness:

1. *Do you know who is on the emergency preparedness committee for your building? All members?*
2. *Do all faculty know the extent and parameters of their role(s) with regard to both natural disasters and threats to the school?*
3. *Do you know who (all) in your building carries a first aid card?*
4. *Do you regularly schedule faculty and student dialogue concerning emergency preparedness?*
5. *If the designated incident commander is unavailable, do you know who is in charge?*
6. *Who is the primary spokesperson in an emergency situation?*
7. *On your school's emergency phone tree, whom will you contact first?*
8. *Do you know where to go on your school website for immediate emergency information?*
9. *What are the means of communication if an emergency occurs outside regular school hours? What communication obligation do all staff members have?*

To determine current knowledge, share these questions as an opening activity at staff meetings about emergency procedure staff development training.

Communication: Students' Role

One of the most important decisions educators can make in emergency planning is the extent to which they include students throughout every phase of the process. Students need to be represented on all emergency preparation committees and must be part of designing the school's emergency communication plan. Further, all students need to understand the importance emergency preparedness content knowledge has in a crisis. Students do understand better when they are involved in the decision making. One final way to ensure students understand the importance of communication during a disaster is through modeling; communicate with families and ask them to participate in creating emergency plans at home.

The more modes of communication used, the fewer children who may be absent or have not fully comprehended the school's emergency communication plan. Students need to fully understand why the emergency communication plan is so important.

Student Home Emergency Plan

FEMA suggests that children encourage families to have a family meeting to prepare a family emergency communication plan. *Why not do the same in the library or classroom?*

Teaching children elements of a good home emergency communication plan may take valuable class time, but in some cases this information can be integrated into other curriculum or content areas. Home emergency preparedness planning should be part of elementary school student health curriculum. For younger students, refer to the FEMA website, where a template for designing a home communication plan for families can be found. FEMA provides a colorful template with standard communication information such as memorizing one's address and phone numbers (for family and neighbors) but also suggests having students provide contact information such as phone number and email for a relative or other close contact who resides out of state and enlisting families in finding a meeting spot should they become separated during a disaster.

A school emergency communication plan differs from a family communication plan in one important factor: It is far less likely that individual students will become separated from the rest of their school or classmates. Therefore, the primary focus for a school's emergency communication plan is that all students understand the plan. They must know where they are to go and what they are to do under critical circumstances.

Communication in Emergency Planning: School Library Curriculum

> The school library may well be the best venue for teaching students about communicating information on any number of topics related to disaster preparedness. Within the context of library and information curriculum, roles the library and librarian can play with regard to communication in emergency preparation include finding information, seeking web resources, determining the best student safety procedures, presenting suggested readings on bullying or violence prevention, learning about home and school safety issues—the possibilities are limitless.

Communication Starts the Recover Process

Ask anyone critically involved in recovery following a major disaster what the most critical issue during and following a disaster is, and they repeatedly respond: "Communication." Communication dynamics following a disaster vary greatly depending on whether the disaster is natural (usually a weather-related disaster) or an act of violence involving the school. With the latter, legal issues come into play, and school personnel usually have gag orders following an act of violence on school property. But students do not.

"We are now dealing with 'Generation Text' instead of 'Generation X,' " said Ken Trump, president of the National School Safety and Security Services. 'The rumors [can] become greater than the issue, problem, or incident itself. Attendance can go down overnight and rumors can fly in minutes" (2014).

Staff should not let children form ideas about what has happened in the event based on what they hear from the media or rumors, or what they pick up from other individuals. Conversely, misinformed students can exacerbate an already distressing situation. Following a major disaster great efforts must be made by all members of the school community to counter the negative impact of counterproductive communication. To help move the school through the recovery process, communicate the honest, simple truth, being cautious to include only what is appropriate for the age of the audience or child.

Chapter 5

Risk Assessment and Threats of School Violence

In the past, proportionately few children in this country experienced first-hand natural disasters or human made violence of a major magnitude. However, rapid changes in society and parallel advances in technology bring disasters and information about disasters into children's lives on a regular basis through personal experience and globally televised media events.

—Zeece, 1998

Assessments, whether looking at the *risks* a school may be exposed to, the *vulnerabilities* to hazards, or the *threat* of or likelihood for a crisis or disaster, are all means for being proactive and better prepared for critical events.

RISK ASSESSMENT

Risk Defined

Terminology in this field is not always standardized, and authors from different institutions and agencies may use the same terms in slightly different ways. For this book, "risk" with regard to natural and human made disasters refers to the "expected losses (lives lost, persons injured, damage to property, and disruption of economic activity) caused by a particular phenomenon over a specified future time period. Risk is a function of the probability of particular occurrences and the losses each would cause" (Coburn, Spence, & Pomonis, 1994).

Psychological Phenomena behind Recognizing Risk

Each school carries its own culture; the cultural phenomenon that defines a school is represented in the make up of its staff, students, and leadership of administration. Characteristics of caring, respect, and safety help define the culture of an individual school.

Education's professional community constantly takes actions to ensure the safety and security of its members. It is in the DNA of educators to do so. At the same time, there will always be certain exposures to risks in schools because at times efforts to reduce them are in conflict with the beliefs and culture of the school.

The same rationale that is used when educators decide not to add bars to the windows of first floor classrooms or install metal detectors for checking in preschoolers is used to make decisions about whether to increase safety measures or how best to ensure a safe environment. Decisions are based around the school's missions and goals. And while these missions and goals are certainly not in contrast with safety and security protocols, some educators suffer an ideological aversion to installing or instilling anything that decreases students' freedom or creates a loss of innocence.

When considering disaster preparedness related to educational time management, it's critical to be as efficient as possible. This is not only because when disaster strikes seconds matter but also because proactive preparation is often relegated to secondary importance and not given critical time priority . . . until it's too late.

An interesting psychological phenomenon exists that the more available the information on any given disaster is, the more likely it is that the event will be *expected* to occur. Often imminence (both geographic and time-wise) of a disaster is directly related to the importance placed on a disaster by the general population—"If it is part of my world, I care"—but the farther away the incident is, the less likely its importance. Other attributes contribute to this "collective memory" as well, such as how dramatic the event is as it unfolds. School safety planners must keep this fact in mind when determining how results are examined and acted on. Knee-jerk reactions to recent headlined events never result in the best or most comprehensive decisions when it comes to disaster preparedness. Such reactions may increase preparedness for the short term, but if nothing happens or the efforts are not sustained, apathy soon returns.

What often happens—it's human nature—when trying to create a safe environment for kids is that recent critical events divert attention from the true task at hand. For example, although fire and flood are the most likely disaster incidents to take place in schools, fire drills suddenly become less important when a natural disaster or a profound school shooting draws national attention. Parents call, asking, *"What are the [tornado, hurricane, etc.] safety measures you have to place to ensure my kids are safe?"* Politicians pressure schools to respond. And suddenly the focus is diverted from the risks that have been identified as real and imminent to what is recent, personal, or immediate.

When officials working on school safety measures focus efforts in areas that have recently made the news rather than on the actual needs of the school, the media is often the reason; highly publicized events increase the general public's likelihood to hyperbolize risk. For example, in general, even well-informed citizens highly overestimate the likelihood of death by plane crash or murder while greatly underestimating the risk of death caused by heart failure or smoking. Why? Because every airplane crash is highly publicized. The librarian, serving as the information specialist for the school, can help members of the school community understand this. Real risks, not perceived risks, are what matter. Understanding the difference between *actual* and *perceived* risk reduces the likelihood of ill-directed emergency team efforts. Risk assessment results can also help alleviate diverting attention from the actual risks that have been solidly identified.

Checklist of Disaster Preparedness: Before, During, and After

The American Red Cross has created checklists for natural and human-caused disasters that are readily downloadable and ready to print in large quantities. The site includes individual checklists related to: chemical emergency, drought, earthquake, fire, flood, flu, food safety, highway safety, hurricane, landslide, pet safety, poisoning, power outage, terrorism, thunderstorm, tornado, tsunami, volcano, water safety, wildfire, and winter storm. They are available in English and eight other languages. Find these helpful checklists at http://www.redcross.org/prepare/disaster-safety-library

Risk Assessment: The Basic Components

Assessing the risks unique to each school may provide a partial solution to the conflict regarding emergency preparedness and keeping students safe while protecting individual freedoms. Risk assessment is productive, positive, and proactive and has the potential to improve response time and efficiency of the response to the disaster.

While it is not the intent of this book to provide comprehensive guidelines for designing school-wide risk assessment, it behooves all members of the school community to understand basic necessary or "best practices" components of risk assessment.

There are generally two major risk assessment foci. One establishes the broader areas of concern around natural, technical, and biological threats. This risk assessment data can guide the direction around which a school focuses its efforts for disaster preparedness. It concerns that for which cause cannot be manipulated, for example, weather-related disasters. A second assessment involves human-induced incidents such as targeted acts of violence. It is the latter, targeted school violence, for which the school plays a more significant role in prevention.

Risk Assessment Perspectives

Most federal funds come with regulatory requirements as to what local agencies must do with distributed allocations. For an entity to qualify for federal funds following a disaster, FEMA requires state and local governments to regularly conduct risk assessments, enact scheduled drills, and do appropriate emergency planning prior to the disaster.

Agencies of civil defense and subsequently emergency services and emergency management have historically been a symbiotic and a shared responsibility of federal, state, and local governments. In Washington State, for example, it took legislative action, passing Substitute Senate Bill 5097 (RCW 28A.320.125) in 2005, to require schools to conduct more than just fire drills in preparation for a broader spectrum of potential disasters, natural or human made.

School districts work under the regulations or other guidance of their state and local government. Most states mandate individual school districts to conduct risk assessments around natural threats and hold emergency preparedness training and drills. (Appendix A provides a sample risk assessment form.) Even if the librarian is not involved with risk assessment, a *well-informed* librarian can contribute on many levels to improved safety standards.

The following stakeholders should be involved in the process: administration, classroom teachers, specialists, custodians, parents, students, community members, mental health professionals, and first responders. A general rule of thumb is that those who will use the plan must be part of developing the plan. The more representation, the more comprehensive the assessment results will be. Members of this team must come together with the same goal of establishing the safest possible educational environment in the school.

High-Risk Areas and Events

One risk at school that should be assessed involves school events. Some scenarios, particularly schoolwide celebratory events, present a greater risk than others. This may be due to high numbers of people, the emotional intensity of the event, or—with regard to school violence—the likelihood of greater visibility of perpetrator. Such high-risk events and areas include:

- *School dances*
- *School athletic events*
- *School political events*
- *Assemblies*
- *School rallies*
- *Large gathering times:*
  - *Dismissal*
  - *Lunch*
- *Large gathering places:*
  - *Library*
  - *Cafeteria*
  - *Parking lot*
  - *Athletic arenas*

Risk Assessment Criteria and Resulting Actions

Risk assessment results consist primarily of the *collection of data* that establishes risk. From these data, current status and areas of need are established through problem identification and analysis. Often these data simply clarify and substantiate what is known or obvious: *We live in Tornado Alley, so that is the natural disaster on which our preparedness exercises will focus.* Some data, however, may uncover areas of concern; new information from which change, training, or intervention must occur; or discoveries for which the establishment of new prevention strategies should be implemented. An example is the post-tornado community effort in Moore, Oklahoma, to provide bicycle safety helmets to all school children.

Risk assessment must be *dynamic* rather than static. While initial risk assessment data (see the sample risk assessment in Appendix A) can establish parameters, some components of assessment are necessarily recurrent, conducted at regularly scheduled intervals, as well as on an as-needed basis, for example, when multiple suicides occur in the school. Still, assessment should be conducted in a *systematic* rather than reactionary manner.

For human-induced critical incidents within the school (e.g., multiple suicides), an honest assessment of the school's *culture* and *climate* can help identify how pervasive key triggers are for such contributing factors as bullying, drugs, sexual harassment, and school violence. Identifying types and frequency of certain behaviors provides focus for crisis prevention professional development goals

As was emphasized in Chapter 4, another essential element of effective risk assessment includes comprehensive *communication of results*. This means sharing results with staff members as well as parents and the broader school community. Students also need to clearly understand risks identified during risk assessment.

Determining Risk

Here are three critical components to consider when determining a school's risk for each identified disaster or incidence:

1. ***Likelihood of occurrence.*** *What is the probability of this disaster occurring in the school community or region?*
2. ***Who and what is impacted.*** *Who and what infrastructure is likely to be impacted by an occurrence of this disaster? For example, with a natural disaster such as an earthquake, every member of the school community has an equal likelihood of being impacted.*
3. ***Elements of risk.*** *What are the ways in which members of the school community and the school building might be impacted by this disaster, that is, what is the degree of loss to each element should a hazard of a given severity occur (Coburn et al., 1994)? For example, impact to the building is far different in an earthquake than when there is an active shooter in the school; with a school shooting elements of risk become far more psychologically complex.*

Sample Risk Assessment: School Parking

In this section, the process of risk assessment is applied to a school's parking situation, one minor region of the school. The questions that comprise this risk assessment demonstrate how complex and specific (yet far-reaching) risks can be found. Questions here address parking concerns but may spark other critical and creative ways to view the school's risks.

Few of a school's structural elements can be readily changed even if risks are determined. Most would require major reconstruction or potentially the need for citizens to vote on the allocation of public funds. One exception is parking. Librarians can work with other members of the school safety team to practice assessing risk by examining the school's parking area(s) and making suggestions for improved safety.

Use the following questions to guide the assessment:

- Do all parking assignments require school vehicle registration and stickers?
- Are student and staff vehicle registration stickers (or other emblems) permanently fixed?
- Are the stickers required to be strategically placed and highly visible?
- Do the registered vehicle permits have reflector features for nighttime visibility?
- Are permits updated annually with each new school year highly visibly displayed?
- If the school has security officers, how frequently do they check vehicles for parking permit stickers?
- Are unassigned or visitor parking spaces supervised and visible?

- How near the school buildings are uninspected vehicles allowed (particularly at high-risk schools)?
- Can the school's employees' assigned parking provide a line of defense?
- Does the school enforce one-way vehicular circulation?
- Does the school's vehicular circulation steer vehicles *away* from classrooms and densely populated areas of the school?
- For multibuilding sites, are parking restrictions ***between*** buildings enforced?
- Are visitor parking areas visible from inside the building for key persons such as security personnel and school administrators?
- Are visitor parking spots monitored by cameras?
- Are visitor parking spaces designated in the most visible and well-lit areas?
- If parking is on the street, do administrator—and other employee—reserved parking places shield the most vulnerable areas of the building?

Simply rearranging or reassigning parking spaces may improve the school's safety and thus reduce risk, particularly in high schools where potential insider risks are greater because student parking exists. Even with something as simple as parking, what constitutes risk at one school may be very different from another school. The application of risk assessment to a school's parking situation demonstrates the need for all members of the school community to contribute with a critical eye to the greatest range of potential risk assessment.

A Question of Risk

When risk assessment data is presented to the school, consider engaging in a dialogue regarding results and provide multiple viewpoints from which school citizens can analyze safety features of the school, or the lack thereof. The following general questions may elicit responses that reveal other ways of eliminating potential risks:

- *How accurate might risk assessment data be on managing this risk?*
- *How likely is this disaster to happen?*
- *Of what potential magnitude might this disaster be?*
- *How frequently might this disaster occur?*
- *How probable are future or additional risks once event has occurred?*
- *How complex might analysis of data gathered from a risk assessment be?*

A BROADER LOOK: WHAT ELSE MAY CONSTITUTE "RISK"?

Disorganization Is a Risk

Some schools may appear to demonstrate great preparedness for various risks, posting plans here and there: *There's a school evacuation plan posted near the door. Every classroom has emergency kits stored . . . In the room somewhere. The list of phone numbers and contact information for local first responders—which may be needed at a moment's notice—is in that big thick three-ring binder labeled "Emergency Preparedness for This School." The numbers for that first responders' contact list are filed in that binder between the list of questions required to respond to a bomb threat and . . . I think it's . . . after . . . it's in there right after the earthquake procedure. And the three-ring binder is around here somewhere.*

If this sounds at all familiar, know that disorganization can be one of the greater individual risks for emergency preparedness for all or any members of the school community. One method of examining risk is to determine how concise, accessible, well-distributed, and well-organized critical information is within the school and within each classroom. "The [risk assessment] committee should combine these bits and pieces, new and old, in a single document, as brief as possible" (Robertson, 2013).

Uninformed and Misinformed

The 2014 Oso Landslide in Washington State showed the tragic results of uninformed and misinformed citizenry. While this region experienced many slides previously, depending on who was interviewed, according to various local media news reports, community members knew . . . or did not know . . . were informed or were not informed . . . of the dangers of living in the slide area. This example is of individual homes, of course. But do not assume it is unlikely that a school may reside in an area of high risk; consider that 110 out of 127 schools in New Orleans were completely destroyed when levy failure followed Hurricane Katrina.

While the degree of destruction caused by any major disaster is difficult to accurately predetermine, for most natural disasters, schools are among the first to be informed of potential disasters. The risk, therefore, lies in what is done with that information. Most communities have completed community threat assessments and hazard mitigation plans through their local department of emergency management that have identified hazards, risk, and vulnerabilities, which are a good source of information for citizens, school districts, and librarians.

School districts, usually from the central office, and various community agencies conduct many types of risk assessments with which teachers may not be involved but should become informed. Particularly when risks are discovered. It is the school citizen's right to know the results of all official risk assessments. Informed is forewarned. Involvement via attending public meetings where these results are shared, such as school board meetings, or reviewing results on the city or county website, can bring previously unknown risks to light.

Mistake in Hearing the Cry "Wolf!"

Considering all tornado activity, an EF-5 super tornado has only a 0.1% chance of happening, and 75% of all tornado warnings are later seen as false alarms. As with the boy who cried wolf, complacency can seep into the community as a result. Such was the case when a devastating tornado in 2011 took 161 lives; people in Joplin, Missouri, had experienced a higher-than-usual number of tornado warnings just prior to this event.

"There's always a possibility you're going to have somebody survive, and people will say, 'Look! There you go! You told us we were going to die, and we didn't!' Then for the next storm, they won't react. Now, people living in high-risk areas will be warned only that they 'may die' if they don't evacuate" (Zaleski, 2009).

Thanks to technological advances in weather forecasting, climatologists have become much more accurate at predicting not only the intensity and location of weather patterns but also human response. As they study weather, they also learn about the sociological impact of predicting and delivering warnings such as the high rate of tornado false alarms, which results from the capricious nature of tornadoes. During emergency preparedness class discussions, therefore, librarians must constantly warn that *any* tornado warning is too dangerous to ever ignore.

The Fiscal Elephant in the Room

Decisions regarding potential risk do factor in the cost in potential lives and injury as well as potential facility expense. Unfortunately, because school districts have limited resources, decisions related to addressing risks—prioritizing and implementing safety improvements—must factor in cost.

Despite their Tornado Alley location, most schools in Oklahoma have neither tornado safe rooms nor underground shelters. During the May, 20, 2013, deadly tornado in Moore, Oklahoma, children and staff at Plaza Towers and Briarwood Elementary Schools "huddle[d] into hallways or crammed into bathrooms or closets" (Murphy, 2014) because the schools did not have storm protection. As often happens, when these two schools—in which seven children lost their lives in that storm—are reconstructed, both will include storm protection.

Oklahoma's politicians wrangle with budgetary priorities as they determine how to address the inadequate facility protection and necessary retrofitting for the 1,100 schools in Oklahoma that need storm protection. "A bill proposed by Democratic state legislators that would've permitted school districts to pay for bond issues out of $1 billion in state taxes levied on corporations never made it out of committee" (Costello and Schecter, 2014), but as fiscal battles rage over whether local school district levies or state budgets should fund protective reconstruction, the schools remain unprotected.

LARGE-SCALE EXTERNAL RISKS

Biological, Chemical, and Radiological Risks Threats

While an incident of bioterrorism—that is, the deliberate release of viruses, bacteria, or toxins that wreak havoc on U.S. citizenry—is far less likely than most other disasters, the possibilities for harm and greater numbers potentially impacted draw attention to this threat (for more information, see the discussion in Chapter 3 on human-made disasters).

Fortunately, to date the United States has not experienced biological, chemical, radiological, or nuclear attacks specifically aimed at schools. Consequently, because so often preparedness *follows* experience, schools may be ill equipped or ill prepared for these potential attacks. Likely, an act of bioterrorism would not be restricted to only the school's population, and federal or outside agencies, trained in bioterrorism response, would be immediately called upon to guide response. It is unlikely that an individual school will be involved in assessing the likelihood of a bioterrorist attack. However, awareness is important; bioterrorism is a growing global concern. Librarians who wish to address the risk of bioterrorism may choose to do so as a hypothetical research project with older students.

Pandemic and Epidemic Risks

An epidemic can rapidly become a pandemic, particularly when an influenza virus develops a new strain. There is no way to assess for risk of spread until a case has been reported, but when risk is discovered, both district-wide and single-school decisions are made with regard to closures due to the spread of disease. Often these decisions are based on characteristics of the strain such as severity, mortality, or attack rate and whether treatment or vaccines are available. The Centers for Disease Control and Prevention (CDC) advocates "social distancing" as one of the most effective means of harnessing the spread of influenza and other contagious conditions. The best way to reduce spread, of course, is through school closure. Prior to closure, ensuring that children observe basic health practices such as hand washing can help prevent the initial spread of disease.

For the most successful prevention measures, school districts must proactively work in collaboration with their local health department to develop a pandemic influenza action plan. The World Health Organization also provides a thorough risk assessment for pandemics which can be accessed online at http://www.who.int/influenza/resources/documents/FluCheck6web.pdf

Environmental Concerns: Mold

Because mold can have such a rapid and profound impact on the library's collection, much has been published on this topic, particularly related to assessment of the library's collection. However, the impact on student and staff health may far outweigh concerns about the library's collection. Mold often occurs as a result of other disasters, particularly those in high humidity regions. District health and safety officials should be called in immediately when evidence of mold is present. Evacuation is likely to result.

SUMMARY

Risk assessment is a critical component of emergency preparedness. While it is unlikely that the school librarian will be heavily involved in conducting risk assessment, by understanding risk assessment; recognizing key triggers and factors involved in risk assessment; being informed of the response process, parameters, and results; and being willing to help communicate information regarding risk can greatly contribute to school safety. Further, recognizing that *risks are everywhere and that addressing risks is a positive and proactive action that improves safety* further protects schools. Informal risk assessment is everyone's responsibility.

THREAT ASSESSMENT

Threat Assessments for Potential Targeted School Violence

> *The origins of human violence are complex. Thinkers, historians, and scientists have explored the issue for centuries, but answers remain elusive. The roots of a violent act are multiple, intricate, and intertwined. The mix of factors varies according to the individual and the circumstances. Understanding violence after it has occurred is difficult enough. Trying to assess a threat and keeping it from being carried out is even more of a challenge.*
>
> —O'Toole, 2008b

For school officials today one of the most frightening and frustrating issues related to school violence is a very clear understanding that there are no absolutes, no clear profile of a potential perpetrator of violence. Active shooters in schools are drawn from a range of student types—the bullies, the bullied, the popular, the loners, the school athletes, the geeks, students with straight A's, students with failing grades, and students with learning disabilities. And while there are certain questionable behaviors about which educators must be aware, professionals warn against using known risk factors to profile certain students as potential perpetrators of violence.

"The use of profiling (i.e., ranking a student's behaviors and risk factors against a set of criteria) is not recommended because it is likely to misidentify youth and in doing so cause more harm than good. Moreover, a process that focuses solely on identification, not intervention, will fail to provide the necessary help to potential offenders" (Jimerson, Brock, & Cowan, 2005).

Instead, it is recommended that when a threat has been launched against some target in the school, school safety officials conduct a **threat assessment**, as suggested in a guide developed by the U.S. Department of Education as part of its Safe Schools Initiative, and use the results from that assessment to help determine the authenticity and likelihood of that threat and then act on it accordingly by applying prevention strategies. In using best practices for comprehensive school safety, school officials focus on the threat itself rather than the individual.

Analyzing the threat also helps remove personality and student history bias. Further, a threat assessment serves to proactively address the need to intervene with potential perpetrators of school violence and focus on seeking help from mental health professionals. Results from the threat assessment may also serve to uncover victims of bullying and other unacceptable behavior carried out on the potential perpetrator. Once threat validity is determined officials intervene to diffuse the threat and prevent attacks.

Six Principles of the Threat Assessment Process
(adapted from the Safe Schools Initiative)

Principle 1: Targeted violence is the end result of an understandable and often discernible process of thinking and behavior.
Principle 2: Targeted violence stems from interaction between the person, the situation, the setting, and the target.
Principle 3: An investigative, skeptical, inquisitive mindset is critical to successful threat assessment.
Principle 4: Effective threat assessment is based on facts, rather than personal characteristics.
Principle 5: An integrated systems approach should guide threat assessment investigations.
Principle 6: The central question of the threat assessment is whether a student poses a threat, not whether a student made a threat.

Historic Context of Threat Assessment

Because of fears and uncertainties concerning school safety that have been created by highly publicized targeted school shootings, institutions and agencies have sought answers as to what can be done to address this crisis. Government agencies have collaborated, spending years studying prior events to determine if there exists best ways to prevent similar tragedies. Since 1998 the Federal Bureau of Investigation (FBI) and the Department of Justice (DOJ) have worked toward this end. They have conducted several studies of high-profile school shootings and published comprehensive results of threat assessment perspectives for use by law-enforcement and school personnel.

The Secret Service (SS), working in collaboration with the U.S. Department of Education (ED), also conducted a study and developed a guide from the findings that resulted in the *Safe School Initiative (SSI)* for individual schools to use when conducting threat assessments. While these reports similarly warn that they are not "a profile of the school shooter or a checklist of danger signs" (O'Toole, 2008b), the results from these major studies bring important findings to light to help us understand targeted violence in schools.

The *SSI* examined 37 high-profile incidents of targeted school shootings to seek preattack perpetrator behavior patterns. The *SSI* built on a formative 1949 study that the Secret Service conducted to determine whether there were patterns of behavior by those who made deliberate attacks on public officials or other persons of renown. This study attempted to determine what, if anything, might have been known about the individual that could have predicted his attack. The Exceptional Case Study Project (ECSP), as the 1949 study was called, concluded that while there were some patterns of behavior that fit into pieces of the broader puzzle, a "profile" of perpetrators did not exist. Instead, the report summarized that a more effective way of addressing threats was to examine the actual threat itself rather than the person making the threat.

While the *SSI* findings reinforced those from the earlier ECSP study as well as corroborating results from the FBI-DOJ study, with regard to some of an attacker's modes of behavior and thought processes or rationales, the *SSI* did identify certain preattack behaviors that show a predilection for targeted violence. "Findings about the pre-attack behaviors of perpetrators of targeted violence validated the 'fact-based' approach of the Threat Assessment process" (Vossekuil et al., 2002) to determine the degree of likelihood the threat will be carried out.

Findings from the Studies

The SS-ED study's identified attackers—that is, those who had successfully carried out targeted school violence—were all males between the ages of 11 and 21, and most had preselected a target with whom they had a grievance. Beyond this, few similarities exist among attackers; half the attackers selected school personnel, half chose students as their targets, and more than half chose more than one target. Despite knowing their targets, in more than half the attacks, other students were killed in these tragic events. Most lived in original two-parent families. Academically, more (41%) were "A" and "B" students than were academically failing (5%). Socially, more were considered mainstream students, though one-third were considered loners by others in the school. These data further demonstrate the understanding that profiling for targeted violence is not effective.

However, some notable facts were uncovered. Most perpetrators had felt bullied, had been diagnosed with a mental disorder, were interested in violence, and had difficulty dealing with loss or failure. Further, in nearly all cases, the attacks were planned in advance, from a few days to even years before. Most attackers had access to and an interest in weapons, primarily guns. And finally, sadly, in nearly all cases other people knew of the attack plans, and attackers demonstrated behaviors prior to the event that caused those observers alarm.

Most often one will never know what actions might have prevented some incidents of school violence, but there exist some key findings from various studies that inform best practices and strategies for helping identify threats, analyzing their content, and seek help for whomever made the threat as well as for the threatened individuals (See the feature in Chapter 3 about active shooters).

A 2008 study built on these previous findings to determine what bystander behaviors could contribute to precautionary measures (Pollack, 2008). Pollack's findings substantiated previous interpretations: It is not only those who threaten or perpetrate school violence who require a trusting adult within the building to address threats; bystanders are also unwilling to report potential threats to anyone other than a responsible adult ally within the school community

> School climate affected whether bystanders came forward with the information related to the threat. . . . In those instances where bystanders with information about a possible attack felt a positive emotional connection with the school or with someone on the staff, they were comfortable coming forward and reporting what they knew. . . . Developing meaningful social and emotional connections with students and creating a climate of mutual respect are essential to keeping school safe. (Pollack, 2008)

Are there implications, then, from these findings that can be used to prevent future attacks? What data can be gathered to help reduce the likelihood of future incidents of targeted school violence?

Conducting a Threat Assessment

The important questions that can be culled from current comprehensive studies to address the prevention of school shootings are: *Could we have known that these attacks were being planned? And if so, what could we have done to prevent these attacks from occurring?* And concerning threat content itself: *How credible is the threat? Is the perpetrator capable of carrying out the threat?*

No one individual can be responsible for determining the degree of authenticity of a threat of violence launched against the school. Assessing a threat requires the school to have a threat assessment team comprised of the broadest possible representation from the school community so that it will produce the greatest range of perspectives and consequently the most thorough results.

Why Librarians Make Great Threat Assessment Team Members

The threat assessment team should have broad representation from across the school and community, yet all members should possess the following skill set and dispositions, which are required of effective contributors (in reading this list it becomes obvious why librarians make excellent members of the threat assessment team).

- *Have an understanding of social, cognitive, and psychological growth and development of children and young adults*
- *Be discrete*
- *Have an evaluative and skeptical yet open-minded mindset*
- *Be fair-minded and trustworthy*
- *Be confidential, that is, able to keep critical information classified*
- *Relate well to all members of the school and larger community*
- *Have training in the acquisition and evaluation of multiple sources of information*

The procedure for conducting a threat assessment (TA) begins when a credible threat has been launched against a member or members of the school community or facility, or a student has acted in a way that can be interpreted as threatening and therefore is cause for concern. The threat may come to the attention of school officials from peers, teachers, or support staff such as bus drivers, monitors, or custodians and may take any number of forms—written or verbal, direct, veiled, or anonymous (for more information, see this chapter's feature about types of threats). Immediacy of initiating the TA process depends on the seriousness of the threat.

Types of Threat

Current school violence threat assessment identifies four distinct types of threats:

- *A* ***direct threat*** *is aimed at a specific individual or group and identifies a specific threatened action, usually addressed in a clear, straightforward manner. If the threat includes violence, it is often described exactly.*

- *An* ***indirect threat*** *may or may not be aimed at a specific individual, but the content and specifics of the violent act (s)—such as time, place, and manner—are ambiguous.*
- *A* ***veiled threat*** *suggests violence but is not explicit.*
- *A* ***conditional threat*** *is often also considered an act of extortion; that is, unless the perpetrator's demands are met, the violence described will occur.*

The TA process involves two steps. The first is the threat assessment *inquiry* that is conducted by the threat assessment team. Not all inquiries will lead to further investigation, but depending on results from this initial inquiry, the second step involves threat assessment *investigation*, which is always carried out by law-enforcement agencies.

The intent of the inquiry process is to determine if a particular student or community resident poses a threat of targeted violence, and it may result in one of several outcomes. Information gathered from the inquiry may lead to the conclusion that no threat exists. But if the threat may be carried out, team members must hand over findings to law-enforcement officials for *investigation* (the next step). Also, if illegal activity is discovered during the TA inquiry, of course, law-enforcement officials must be informed.

The scope of data gathered during the inquiry is determined by the degree of seriousness. For example, if the threat is posed at a specific imminent time, team members may need to gather immediately and take action.

When a threat has been discovered or launched, team members should begin gathering data. Questions concerning the threat may include:

- Who made the threat?
- Under what situation?
- To whom or what?
- In what form?
- What is the motive for the threat?
- Has this person made threats before?
- Has this person committed prior acts of violence?
- How specific is the threat?
- Were there witnesses?
- When is the threatened action to be carried out?

Inquiries need to first gather identifying information about the threatener such as name and description. Additional pertinent personal information may provide clues and/or red flags such as friends and social connections, school performance, history of bullying or being bullied, and connection with trusted adults in the school. Inquiries look for red flags such as a history of violence or anger management issues, obsession with weapons, access to weapons, substance abuse, and criminal records. Finally, current situational information such as a romantic breakup, change in academic performance, change in home or other family situation, or a recent loss, grudge, or conflict with another school member may represent key triggers.

With a recommendation that the school librarian join the threat assessment team, the information in this chapter is merely initial informative content meant to introduce and recommend the use of threat assessment; it does not constitute actual training. Whether or not the librarian volunteers to become a member of the threat assessment team, learning about the impact creating a culture of respect, safety, and caring has on the school's safety, as well as learning about threat analysis, contributes to the body of knowledge a school needs when threats occur.

Threat assessment team members must seek a motive from the data they collect as well as from data they collect through the interview process, which usually begins by interviewing students and adults who know the student of concern. Interviews must center on the facts and ask those interviewed to remove emotion and opinions as much as possible. Interviews with members of the family must also be conducted, with the awareness that family members may try to protect as well as conceal personal family information.

Therefore, during family interviews emphasis must focus on seeking help for the threatening student. Finally, if the threat is launched against a student or students, those threatened need to be questioned. A student has a right to know he or she has been threatened; after informing the student of the threat against him or her, during the interview questions should center on relationships, and queries of motive should be addressed.

Finally, the student who made the threat must be interviewed. He may expect legal representation or family members present. The primary purpose of this interview is to diffuse a potential attack; therefore, the tone of the questioning should send the message that the threat does not have to be carried out, alternatives to violence exist, and help is available. In any case, the message must be sent that threats are taken seriously.

Red Flags That May Accompany or Indicate a Threat

- *Seeking weapons*
- *Inquiring about bomb building or searching online*
- *Talking of suicide*
- *Showing heightened interest in previous school attacks and violence*

One critical cue to look for is whether the student, when interviewed or in his initial threat, reveals *very specific details* concerning the threat. For example, it is a call for alarm if the exact place, time, victim(s), and how the targeted act of violence will play out has been described.

After assessment information has been gathered, threat assessment team members must determine a course of action. Analysis of the threat assessment may lead to the conclusion that the student does not pose an imminent threat but should be closely observed or referred for counseling, or the results may lead to the conclusion that the threat is not authentic. After examining interview results, if the TA team believes the threat is real, they may need to turn the threat over to law-enforcement officials for further investigation.

FINAL NOTE

It is important to note that the threat assessment focuses largely on threats from the immediate school population, most often students within the school. That is, in the case of an outside active shooter entering the school building such as happened in Sandy Hook, it is unlikely that necessary information can be collected to provide adequate analyzable threat data prior to an attack. Threat assessments are more effective with in-school populations that are known members of the school community and from whom pertinent data is readily available.

STUDENT ANONYMITY WHEN REPORTING THREATS

Breaking the "code of silence" present in so many schools requires all students to understand that reporting threats is critical to everyone's safety. Establishing a safe environment for reporting requires anonymity. "Students must know their reporting is anonymous," John McDonald (personal interview, 2014), executive director of security and emergency management at Jeffco Public Schools in Golden, Colorado, said as he explained the success of Jeffco's safe schools.

One of the most effective methods employed for preventing school violence is to establish a hotline by which students can anonymously tell authorities of threats. Safe2Tell and Safe2Text (personal interview, McDonald, 2014) are two dynamic tools that have been effectively used to report and prevent many potential assaults.

Why Building-Specific Crisis Planning and Assessment?

By now you may be thinking, "My school has a district-wide plan in place for dealing with school violence, emergency preparedness, and natural disasters that was created by the district safety office or department. Why should I do more?" This list, by no means complete, offers answers:

- *Because every building is unique. Unless your school district has devised a specific building-by-building response plans for crises such as an active shooter and each building has been assessed and is response ready, the likelihood is slim that your building is safe.*
- *Because everyone needs to know. Unless ALL your staff have been trained in preparation for an active shooter—that is, knowing exactly what to do, where to go in the school, wherever they may be—the likelihood that staff are response prepared for what to do specific to your school is unlikely.*
- *Because every student population is different. You know your student population better than anyone at the central office.*
- *Because every individual school has its own culture. There may be issues to resolve, students who will report and otherwise help, and unique leadership within the culture of each school.*
- *Because an individual school is more able to nimbly plan, respond, drill, and assess emergency needs than an entire district. Therefore, response time is reduced.*
- *Because parents and other family members need know.*
- *Because top-down mandates are less likely to be endorsed and followed than when people in the school are involved in the decision.*

Report All Threats

Threats happen. All members of the school community—staff and students alike—must take threats seriously. Reporting a threat does not necessarily lead to an accusation to the threatener or a liability for whomever reports. Regarding threats, consider these important points:

- *Ignore nothing. If it sounds like a threat, it is a threat.*
- *Report threats immediately; do not wait, ponder, or worry about whether a threat is real.*
- *Do not try to interpret or analyze the threat.*
- *Write it down. Immediately upon hearing a threat, write down exactly what is heard or remembered. Try to include exact wording and facts of the threat as well as time and date; as much as possible, remove emotions and personal reactions to the threat.*
- *List witnesses.*

Threat Assessment Is Only One Tool

Examining the school's culture and ensuring a climate of respect is paramount to school safety. That is, while threats will occur, the most effective preventative measure comes from the school's own established culture and climate. Threat assessments are intended as only one measure the school may use to reduce violence.

> *The threat assessment process by itself is unlikely to have a lasting effect on the problem of targeted school violence unless that process is implemented in the larger context of strategies to ensure that schools offer their students a safe and secure learning environment. The principle objective of violence reduction strategies should be to create cultures and climates of safety, respect, and emotional support. (Vossekuil et al., 2002)*

Use guidelines for threat assessment in the school as a means of establishing a process to investigate, evaluate, and mitigate targeted school violence.

Prime Prevention Ingredient: A Culture of Respect

Establishing a culture of respect and a climate of safety is the most important ingredient in preventing school violence; all students must feel valued, recognized, respected, and heard.

> Connection through human relationships is a central component of a culture of safety and respect. This connection is the critical emotional glue [between] students, and between students and adults charged with meeting students' educational, social, emotional, and safety needs.
>
> In the climate of safety, students have a positive connection to at least one adult in authority. Each student feels there's one adult to whom he or she can turn for support and advice if things get tough, and with whom the student could share his or her concerns openly and without fear of shame or reprisal. (Fein et al., 2002)

Studies of targeted school violence have found that a key element in establishing a culture of respect and safety is that students must have at least one responsible adult with whom they can connect. Most students who become active shooters within their own school have told at least one fellow student of their plan, but those plans are never shared with a responsible adult in the building. Pollack (2008) found that student bystanders who experienced attacks in their school but had not reported recognized threat(s) reported that they could not identify that one trusted adult in the school with whom they felt safe in sharing their knowledge of the threat. One of the primary objectives, then, for preventing school violence is to establish an environment in which all students feel safe in sharing concerns, serious threats, and potential incidents of violence.

Faculty meetings often happen in the library. One contribution the librarian can make in promoting a culture of safety and respect is to begin each faculty meeting by proactively discussing students of concern and asking faculty members to "adopt" a student with whom there's a concern. If the school is not willing to take on this adoption policy, librarians may connect with students of concern through literature, research, or technology. Showing an interest in students' literary taste or research interests can open doors, connecting staff to students in distress.

Safe Schools "Climate Control" Features

- *Emphasizing critical and deep listening to students*
- *Breaking the "code of silence" around reporting threats and incidents*
- *Having zero tolerance and an antibullying campaign, school wide*
- *Ensuring every student is connected with at least one responsible adult within the school community*

SUMMARY

While risk assessment results provide a comprehensive approach to safety with the purpose of mitigating potential hazards, the threat assessment, specifically that concerning targeted school violence, examines events as they happen and is conducted at a more granular level. Both processes are critical to school safety. Although comprehensive risk and threat assessments may be beyond the purview of the school librarian, the librarian's understanding the roles, dynamics, and significance of each of these components contributes to the broader school community by providing one more well-informed, safety-conscious individual within the school community.

Chapter 6

Vulnerability Assessment for the School Library and Beyond

Vulnerability assessment is a sort of "self-examination" process whereby schools seek out, identify, and evaluate potential areas of weakness capable of impacting safety. The rationale and goal for vulnerability assessment is the sustained safety, security, and health of all members of the school community.

We have all heard and applied the timeless saying that "The best offense is a strong defense." Rarely is there a more fitting application for this age-old adage than when preparing for disasters—both natural and human-made. Preparation involves examining the library, the school, and the greater surrounding community for issues of safety, stability, and durability. Being proactive with regard to vulnerabilities will protect property and save money and lives.

Vulnerability Assessment Team

Like the broader emergency preparedness committee, the more members of the school community are involved in assessing the school's vulnerabilities, the greater the security and safety results. Consider asking parents who are known medical professionals; first responders such as firefighters, paramedics, and police officers; or those involved in safety careers to help conduct the school's initial vulnerability assessment. Some professional parents may find this to be the one school volunteer committee or activity for which they feel well equipped to contribute and valued for their expertise. These local experts can then pass off their results to their respective agencies, further involving the community in school safety. Also include representation for students with access and functional needs as well as students who are learning English; perhaps a parent, a local community member, the special education teacher, or a student representative would work well in this position. If this request is a one-time commitment, it is likely to draw greater involvement by parents.

The vulnerability assessment will vary from one school to the next depending on budget, student population, school culture, and the building's physical structure, size, and location. While vulnerabilities unique to the school library exist beyond standard classroom factors, there are some universal considerations that every school should evaluate.

The vulnerability assessment must be viewed through two different lenses: potential natural hazards and disasters from human threats. While some overlap exists, each potential disaster carries unique characteristics to consider, for example, factors to consider regarding a potential active shooter in the building are very different from considerations related to flooding or fire.

Unfortunately, checking for vulnerabilities is a step that is often overlooked in the most conscientiously well-prepared community. And while schools and communities may stockpile water, prepare survival kits, and even have faculty enroll in safety and first aid courses, injuries and losses can be minimized and some disasters prevented altogether if the school and surrounding community first take steps to evaluate potential vulnerabilities.

An excellent source of information for the vulnerability assessment team is the local office of emergency management. Most communities have done these assessments for both manmade and natural hazards, and the local department of emergency management can provide this information (some even post it on their websites).

PHYSICAL VULNERABILITIES IN THE SCHOOL LIBRARY

The librarian who volunteers to be part of a school's vulnerability assessment team should chart findings to evaluate the school and surrounding community. Some districts have created their own vulnerability assessment chart. The U.S. Department of Education also provides a free 50-page guide to assessing school vulnerabilities called *A Guide to School Vulnerability Assessment*, which can be found on their website at http://rems.ed.gov/docs/VA_Report_2008.pdf

Regardless of which assessment instrument is used, all research directs team members to ensure that vulnerability assessment is an ongoing process. The team should enlist information produced from the initial assessment to make recommendations for mitigation measures, improvements, or changes to the school via the school board or responsible governing body.

If the librarian is not part of the all-school vulnerability assessment team, he or she should at minimum conduct a vulnerability assessment of the school library. The way to find and correct vulnerabilities is to conduct walk-throughs and physically check the library for areas of concern. The following vulnerabilities are universal to most school libraries. In addition to the vulnerabilities listed here, each school library has vulnerabilities that are determined by each facility's unique features.

Shelving

Yes, librarians want the books to fly off the shelves but certainly not in a disaster. First, check shelves for vulnerabilities. Libraries should not shelve books on the top shelves, or those books should be secured. Bookshelves should not be close to primary seating areas. Although most bookshelves remain stationary, rearranging other furnishings even slightly, particularly in the primary seating areas, may improve the room's safety. The shelves must be secured if they are not already so that they do not fall over and hurt students or staff or block exits. To determine this shelving safety issue, visually imagine the room shaking: Where could books and other material land?

If the school has any possibility of flooding, books should be shelved at a minimum of four to six inches off the floor. Even in schools where no natural flooding is likely to occur, if the school library is on the first floor or in the basement, the potential for water damage from water pipes bursting must be considered.

Every library is designed differently, so it is difficult to establish specific standards for how shelves should be arranged. Thus, use this simple rule of thumb: To improve shelving safety, imagine where students are likely to be should a disaster strike and then examine all potentialities for toppling, falling objects, or other safety issues. Make necessary changes based on these observations.

PERSONAL STORY: 2001 Northwest Earthquake

On a cold mid-February day in 2001 a sizable 6.8 magnitude earthquake shook the Pacific Northwest, a quake thereafter known as Nisqually Earthquake because of its geographic epicenter. I remember initial panic while rapidly forcing students under tables before taking cover myself; the drills we had practiced previously performed like clockwork.

After standing hours in the freezing cold, we were allowed inside. Students returned to class; I went alone into the library. The displays on the six-foot tall shelves were leveled. Nearly all the books had fallen either onto the floor or onto the shelf. It was, fortunately, a non-injury lesson regarding that seemingly innocent choice for placement of books on display.

—Christie Kaaland

Televisions and Other Large Devices

Large televisions on moveable stands often with other accompanying viewing equipment such as a DVD player should be well secured, usually bolted to stands, to prevent toppling during an earthquake or other disaster.

> Some librarians are charged with the building's technologies and equipment checkout and accountability. As important as knowing how many TVs and computers the school owns is knowing that all such devices are secure on stands and shelves. If teachers share this equipment and portability is a necessity such that bolting down is not possible, ensure that stands or carts are not so top-heavy that they can tumble or violently move around and injure students or staff in the event of a natural disaster.

Windows and Lights

The librarian's personal preferences, pedagogical leanings, and creative space style influence the arrangement of library space. Each library is unique in its displays and room arrangement, including how much light fills the spaces. Even if windows are uncovered every day, window coverings are now a necessity. During a lockdown or lockout, windows must be covered, and it needs to happen quickly. Ensure that visual exposure from the outside of the building and, internally, visibility to the rest of the school, can be obscured so that students are not visible to anyone outside the library facility.

PERSONAL STORY: Library Lighting

My personal preference: I love light, and the brighter the better. The last school library position I held had one entire wall of north facing windows. Much to the frustration of the custodian, I never allowed the curtains to be drawn during the school day, nor did I close them at night. I believe children need more natural light than they are often afforded during the school day, and I felt that the brighter lights help protect their eyes while they are reading. However, by today's disaster preparation planning standard, my beliefs concerning window covers revealed unwise safety issues. If I were in that same school today I would, at minimum, have a plan for quickly covering the windows to protect students from outside visibility and investigate clear earthquake film that would prevent the windows from breaking due to ground shaking.

—Christie Kaaland

Signage

Signage for emergency planning at minimum should include signs for at least two appropriate exits from designated locations. Consider using signage as a way of informing and reminding students of emergency procedures and safety factors. Additional signage can be a constant visual reminder to the school community of the importance of vigilant preparedness. As part of the vulnerability check, look for emergency exiting signage in each classroom and hallways, and check that exit plans for each locale use the most current school maps.

Displays and Collectibles in the Library

Probably the most common vulnerable consideration is that of the displays in the library. It is one that can be easily managed, it may be obvious to outside observers, and yet it is often not recognized as a vulnerability by the librarian who observes it daily and is perhaps guilty of creating a safety hazard. This may take a critical eye. Survey the library to make sure that students experience a safe place with regard to all displays.

> Some libraries are so filled with knickknacks, souvenirs, displays, and other trappings that visitors can scarcely breathe. While a collection of reading buddies such as stuffed animals and other objects of literary reference can provide important bibliotherapy for patrons, nothing is more important than student safety. So whether it is a personal collection of literary bears, an academic collections of primary sources of local history, or proudly displayed student work, make sure these collections in the library do not interfere with student safety as potential falling objects or exit blockers.

For librarians who are collectors, the most important question to ask may be this: *What displayed objects are appropriate for the library with regard to patron safety?*

What Do We Do with Dewey: Live "Collectibles" in the Library

Few librarians have not heard the true story of the stray cat Dewey, who lived in the Spencer, Iowa, public library. In some libraries live animals permanently reside and provide an engaging opportunity for children who love animals and for those who do not have pets. It is not suggested this wonderful opportunity be excluded from the library. However, always be aware that these animals rely on human care and are vulnerable during a disaster.

> If animal friends reside in the library, make sure there is a plan for their safety, care, nourishment, and residence should the library experience emergency closure. Depending on both the types of animal as well as the disaster potentialities for the school library, have a plan *and a backup plan* for the care and feeding of library pets in case of a disaster or closure.
>
> Animal behavior becomes unpredictable during a crisis. Even the most gentle classroom pet possesses primary survival instincts that may show up during an emergency, particularly during a natural disaster such as an earthquake. In the unlikely event that a library pet could potentially cause harm to children in the library, such as scratches or bites, make sure their living facility is secure enough to withstand an earth shaking disaster.
>
> In addition to animals, plants can provide a safety hazard. When procuring plants for the library, know their "story." Make sure the plants produce no health hazard, all seasons, and have not had deadly pesticides applied. Make sure they are secured so that they do not create a falling object or exit blocking hazard. The best choice is to research plants before purchasing them for the library. Organic air-purifying plants are a good option.

Electrical Outlets and Multiple Connections

As technology needs rapidly increase, particularly in older buildings that have not been electrically retrofitted, libraries and classrooms often take electrical risks related to powering up computers and other technology. Each outlet and power strip has a recommended maximum amp capacity. Check connections and do not exceed the recommended capacity.

Even without the chaos of a disaster, safety issues regarding connectivity should be checked regularly. To prevent tripping ensure power extension cords do not lie in foot traffic areas. Librarians must look for extension overload, wires running across open floors, and use of multiple extension cords.

Visibility

Some schools, particularly high schools, are fortunate enough to have very large and spacious school libraries. But there may be areas of the library completely out of view and even out of earshot of the school's loudspeakers. Check around the library to make sure there are not corners, furnishings, or displays that can block patrons' view such that they may be excluded from all-class emergency actions. If changes cannot be made, make a mental note and place a note in the substitute folder to ensure all students are present and accounted for when emergency plans are set in motion.

Capacity

What is the capacity of the school library? On the library's busiest day, if a school lockdown or shelter-in-place order occurs, can the facility accommodate the number of students present for an extended period of time? Although little can be done other than on a large-scale structural basis with capacity vulnerabilities, awareness may help curtail potential harm. Being aware of capacity issues in other parts of the school may help protect children when emergencies arise.

Another capacity issue involves the speed by which students can exit the building. Vulnerabilities in this area can be easily discovered through emergency drills. Practice drills (see the discussion in Chapter 7 about drills) with different entrances and exits from the building blocked to discover exit capacity issues.

Exits, Entrances, Doors, and Locks

The library, like every room in the school where students go, should have a school map posted that identifies a minimum of two exits (see the section titled "Signage" earlier in this chapter). In schools with open floor plans, exit paths should be obvious and kept free of obstruction.

Another simple safety factor to consider involves the internal locks on the library doors. Nationwide, one decision many schools made following the Sandy Hook shooting was to replace classroom door locks with locks that can be secured from the inside. In the event of a potential active shooter, librarians should not have to step outside the library to secure students inside. While some schools require that classrooms maintain locked doors throughout the school day, locked doors can actually be a safety hazard. Some traffic flow is necessary.

Standards vary as to whether classrooms stay locked or remain unlocked during the school day or when the school is open for activities, but most classrooms are unlocked during the course of the school day. Some schools chain external school or gym doors to eliminate outside intruders. While such well-intended acts may protect students from external intrusion, imagine having to remove such cumbersome chains if a mass exit is necessary, such as with a fire. Most important, you need to be able to easily and quickly lock and unlock library and classroom doors from the inside.

An active library invites movement, and this includes furnishings; children physically move chairs and tables to arrange better research group dynamics, and teachers and staff rearrange the room to collaborate at after school functions. Librarians can help everyone keep safety in mind as library visitors change the dynamics of the room with just a few simple changes in arrangement decisions.

Movable objects in the library can easily and unknowingly block exits, so librarians must continually check exits to ensure they remain unblocked. School library exits should remain unobstructed when school is in session or the library is in use.

Physical Vulnerabilities in Other Classrooms and Beyond

Librarians may wish to share the results of their library's physical vulnerability assessment with others in the school whose classrooms are known to be vulnerable. Are there classrooms in the school in which teachers live in a maze of cluttered material? Sharing the library vulnerability assessment results at faculty meetings may incite further action within the school building.

Other vulnerabilities exist beyond the classroom such as equipment in the gymnasium or on the playground. Most school districts have occasional safety checks, and they are often conducted across the district, community, or even city. "School assessments should not happen in isolation but instead should occur in the context of community assessments and in coordination with community partners, which can build a community wide commitment to safer schools" (U.S. Dept. of Education, 2008). Check with the school district's safety office to determine the district's vulnerability assessment plan. As mentioned earlier, also check with the local department of emergency management for information about specific community vulnerabilities.

No educator who has experienced a major disaster can deny that such safety checks happen all too infrequently.

TECHNOLOGY AND DATA BACKUP

The School Library Website

Although few school websites today provide access to students' private information such as address, phone numbers, or emergency contact, how easily someone can access this information and release student data can be determined by a brief school website check.

Though the librarian should not assume responsibility for other school website content, simply pointing out vulnerabilities to school officials may help maintain students' safety. In the library make sure the public cannot access student records via the website. Even something as seemingly innocent as checkout records should not be accessible online to anyone other than the patron.

As appealing as they may be and as important as the library website is to its virtual presence, extra caution should be used prior to posting student photos on the school library website. Districts require a parent photo release authorization prior to these postings. Librarians should take extra caution, however, and ensure that parents understand the omnipresence and risk when granting permission to post student photos. For students' safety, photos of any child whose family has custody or other personal issues should simply not be posted on the school library website. In sensitivity and fairness to students with such issues, post instead students' exemplary work.

Library Inventory of Assets

One of the most valuable actions school librarian Laura Pearle (2014) took prior to the fire at Hackley School was conducting her annual school library inventory, printing out results of that inventory and holding them at a second location. Consequently, Pearle was able to produce for the insurance company a comprehensive inventory of all library material so that she could replace all the material. Off-site backups can provide needed critical data and, more important, reduce the school librarian's stress following the destruction of all or part of the school library.

Public Access Catalog (PAC) vendors can also help in both backing up data off-site as well as reproducing needed data following a disaster. The Library of Congress (http://www.loc.gov) and the American Library Association (http://www.ala.org) offer additional resources for material reparation. In fact, several books have been published (Alire, 2000; Matthews & Feathers, 2003; Fortson, 1991; Kahn, 2012; Wellheiser, Scott, & Barton, 2002), and many web resources provide excellent resources for recovery and repair of lost data, material, and equipment.

SUPPLY VULNERABILITIES

Batteries Save Lives

Batteries are one of the most critical supplies in an emergency situation, particularly when students are forced to stay at the school for long periods of time such as during a shelter-in-place situation. The school must have a large supply of all types of batteries. While battery inventory needs vary depending on use and school size, at minimum, a large supply of AA, AAA, C, and D batteries is a safety standard. Purchasing additional batteries specific to individual technology and equipment is wise.

One of the simplest safety checks a librarian can perform is regularly checking the battery supply. Ordering batteries only once a year can be shortsighted; batteries surprisingly disappear, and supplies run low unbeknownst to the librarian. Check battery supplies frequently and refill when necessary. Match battery needs with technology and other equipment that may be needed or used only during an emergency such as walkie-talkies. Consider holding some batteries separate from the school's normal battery supply, labeling them for and storing them in emergency kits. Kahn (2012) suggests keeping a battery checklist that includes battery location, serial numbers, and hours of power.

Depending upon budgetary constraints, the librarian who is fortunate enough to have a school safety budget should also consider purchasing hand held cell phone battery chargers. The most critical communication tool during a disaster is the cell phone and the primary mistake made following a disaster is the careless overuse rather than preservation of cell phone charges.

COMMUNICATION VULNERABILITIES

PERSONAL STORY: Isolation

During my first lockdown experience seven years ago, I was a special education teacher teaching in an isolated, separate portable building. We were given instructions over the intercom to lock down but with no additional information. My students and I could hear helicopters overhead, but we had no idea what was going on outside the portable or even in the main school building. Today so many people have cell phones, but I didn't have one then. I was fortunate to hear from my son. He called me on my classroom phone to tell me a robbery had taken place nearby and that the perpetrator had not been apprehended and was potentially near the school. The most frightening part of the experience was being in a vulnerable and isolated portable with no communication from the outside. We had no idea what was happening.

—Metcalf, Feb. 28, 2014

Though it is frequently overlooked, one of the simplest vulnerabilities to address relates to the school's communication plan. A disaster preparedness plan, *including all communication components,* is effective only if all members of the school community know about and understand the plan. In a 2006 study 95 percent of school districts reported having a disaster plan in place, yet less than a third reported having practiced any part of it, including the simple act of practicing lines of communication.

In large-scale natural disasters such as hurricanes and tornadoes, first responders consider communication to be the most critical factor related to response and recovery. And this is even more important in the event of human-caused disasters such as school shootings. Most first responders could share stories—both successes and tragic failures—that attest to the importance of establishing clear, concise, pre-planned, practiced, and research-based lines of communication in disaster preparedness, response, and recovery.

Releasing Authority

Just before classes officially resumed following winter break the morning of January 2, 2007, twenty-year-old Douglas Chanthabouly easily entered a school building, walked up to 17-year-old Samnang Kok, and fired three shots close range, killing him instantly. What followed began the shortened stint in office of the new Tacoma School District superintendent. Although at the time the district had an established system of jurisdiction and communication, the relatively new school superintendent, a retired career marine, wrangled with local police over authoritative decision making at the scene. The confusion regarding authority and communication that followed was given media attention alongside the tragic murder. Though this incident happened inside a school building, as most school district policies would acknowledge, the severity of the action required police intervention. The importance of immediately and cooperatively relinquishing authority when criminal activity reaches such severity as a school shooting is critical.

Librarians involved in helping disseminate information should consider the following six best practices with regard to an emergency communication plan. These are in no particular order but rather depend on each school's current communication plan (or lack thereof) and needs.

Consideration #1: ***Does the district or school currently have an emergency communication plan as part of its emergency and disaster plan(s)?***

If the district has a *district-wide* emergency communication plan, the librarian may simply lend a new voice and perspective in a welcome review of the current district plan to implement a *schoolwide* plan from its results. A quick inspection of this emergency communication plan should establish its currency with regard to the following factors:

- Changes in personnel, particularly administration
- How recently all data and contacts have been updated
- How well the plan has been communicated to all adults in authority in the school community

One simple but critical emergency planning leadership role the librarian can take is to volunteer to keep the emergency communication plan contacts updated annually or as personnel changes occur. Check also that the emergency communication plan is clearly labeled with two important and distinct dates: the date that the emergency communication plan was created and, more important, how recently it has been updated. If these dates have inadvertently been omitted, suggest to administration that they be added to the documents and updated as changes are made.

Consideration #2: ***How recently and how well has the emergency communication plan been distributed?***

Whether the district has established district-wide standards for every school's communication plan or the individual schools created their own, comprehensive distribution of the plan is critical. The plan should be widely *and frequently* distributed so that it is readily available, in print and online, to all. Once the school or district has a well-established and well-designed communication plan that includes an emergency contact list, the librarian can work with administration on the next steps needed to address vulnerabilities. The important next step involves dissemination of the communication plans and procedures.

Consideration #3: ***Does the school have a plan for updates, changes, and reminders for the emergency communication plan?***

One simple suggestion the librarian can make to help ensure the continual dissemination of information on disaster preparedness in general, including changes and updates with regard to minor changes, is to request that disaster planning and emergency preparedness be an ongoing item on the agenda of every faculty and PTA meeting. Here, the librarian can assume a disaster preparedness leadership role. As the school's information specialist, the librarian can disseminate new national disaster communication updates or current research regarding disaster planning" An ongoing meeting agenda item is also a constant reminder to the school community to conduct class emergency drills and regularly discuss with students the importance of emergency preparedness.

Communication Vulnerability Considerations

In establishing the next important steps in checking for communication vulnerabilities, consider addressing the following questions:

- *First and foremost, with regard to emergencies, do all adult employees in the building know who is in charge?*
- *Do all staff members and other adults in the building know whom to contact first in cases of different disasters?*
- *Do all staff members have the emergency contact list readily available? Are the emergency contact list and communication plan part of the teachers' established classroom routine, including substitute folders?*
- *Has the emergency contact information been disseminated to parents as well as the parent teacher organization (PTA)?*

Consideration #4: ***Are there gaps in the communication plan?***

Gaps happen.

Ironically, one of the most common errors in the communication planning phase is the exclusion of students. And students need to know. When considering the overall "vulnerability" in vulnerability assessment, certainly the students in the school are the most vulnerable of all and must be considered first.

Make sure there are no physical communication gaps. Just as Lisa Metcalf (March 21, 2014), currently a librarian and previously a special education teacher, pointed out, communication with certain locations of the school can be overlooked. When communication fails, not only do students endure undue stress, but potentially lives are at risk.

Consideration #5: ***Is the school's emergency communication plan distributed*** **officially?**

Even if the emergency communication plan has been updated and well designed, placing it in teachers' boxes is not an effective way to ensure that all members of the school community will read, post, and become familiar with the plan. What will happen? Some teachers will carefully read and understand changes, while other teachers—busy with the day-to-day running of their classes—may not even remember to replace earlier plans. What happens then in the case of an emergency? Some teachers may respond based on a previous emergency communication plan! This supports having disaster planning as a regular item on the school's faculty meeting agenda. Instead of placing an updated emergency communication plan in teachers' boxes, distribute it at faculty meetings and spend time going over the communication plan in detail.

Consideration #6: ***How are decisions regarding communication made? Avoid unwarranted and knee-jerk changes.***

Like so many factors of disaster preparedness, plans are often created and established immediately following a disaster that has been given a lot of media attention. *Any* parts of disaster planning that are knee-jerk responses to recent events will likely have gaps. Changing a communication plan in response to an attention-grabbing disaster often leads to a narrow focus on a single issue that was brought to the public's attention through the media rather than comprehensive reevaluation.

Instead, make such a time a teachable moment. Such cases provide excellent opportunities to promote comprehensive, valid, authentic information. Authenticating and validating information is at the heart of library work.

PEOPLE VULNERABILITIES

While schools have established policies for signing in volunteers and visitors, it is impossible for office personnel to identify every person who enters the building. Until recently, strangers could easily and confidently enter a building and sign in as an invited guest.

Adding simple steps to a school's sign-in procedure may stop uninvited people from entering the building. Depending on the school's unique community and cultural characteristics, size, and grade levels, schools may want to increase their visitor security policy by:

- Identification badges with photo ID for frequent visitors or volunteers decrease office congestion and unnecessary questioning.
- For single special events in which family members are invited into the building, include forms sent home with students that family members carry back to the school as their "invitation" to attend.
- All school community members, including classroom teachers, should have coded security questions that are changed regularly.
- Librarians and other specialists who see every child in the building should obtain a list of any potential child safety concerns (e.g., volatile noncustodial parents or guardians).

Suggested School Security Questions

For all security questions, decide on a designated answer, number, or fact. Whether or not it is a correct answer is less important than knowing the correct response under a crisis situation. Answers should be simple, preferably one word. Share these answers with appropriate staff members. Change security questions at least twice per school year.

All-School Employee Security Questions:

What is the population of the school?
Who is the oldest person on staff?
Who was the school's first principal?
What is the principal's first name?
What year was the school's first graduating class?
Who is the current school board president (or the president's spouse)?
What is the school librarian's favorite wine (or other beverage)?

Individual Staff Member or Librarian Security Questions:

What is the oldest book in the library?
What is the librarian's alma mater?
Who was the school's first librarian?
What is the principal's favorite dessert?

Security Questions Related to Informing the School about an Armed Intruder:

(Coded responses should be established, for example, to the first question below, "Yes" means there's an intruder in the classroom, "No" means there's not.)
Will you be taking your class to the Team Red Meeting today?
Where is the after school track team meeting? (for location alert)
What is the school's internet password? (assuming the district uses best practices and changes the schools' and district's Internet password annually; various answers provide coded information such as location of intruder)

VULNERABILITY OF AND TO STUDENTS WITHIN THE SCHOOL

Although information related to school violence prevention is covered in other chapters, here is a reminder that crises and emergencies come from both external and internal sources. Every school has the potential for student acts of violence. The librarian, one of the few people in the school who sees every child over many years, may be one of the few adults who notices students' behavior over the long term, including changes in behavior. The librarian is neither a counselor nor a mental health expert. However, reporting suspicious actions to those experts may help a student in need or a school under threat. Reporting such observations to the school counselor or mental health personnel may be simply an informal suggestion to the school counselor or an official report.

Protecting the Most Vulnerable

A vulnerability assessment means it needs to protect the *most* vulnerable. Ensure that the following considerations been addressed:

- Will all students with disabilities have equal protection?
- What arrangements have been made for ensuring the safety of all students based on individual diverse needs?
- Is there preparation in place to help students who do not speak English to clearly understand dire circumstances?

Vulnerability Time Line and Documentation

To ensure that weaknesses and dangers discovered during vulnerability checks are acted upon, it is advisable to create a time line of actions and results.

Following the vulnerability assessment a plan of action for addressing hazards should be constructed. Prioritize the greatest vulnerabilities, that is, those that pose the most imminent risk to students and staff.

One important feature that a time line can provide is documentation, for example, information needed for potential legal considerations as well as reminders of what has been discovered and what has and has not been addressed. This paper trail also provides sound evidence for requesting personnel and fiscal support from school boards or other governing bodies.

Vigilance is key.

Two final points:

1. The greater the number of responsible stakeholders involved in checking for vulnerabilities, the more likely potential challenges and dangers will be discovered and corrected. Remember, any part of checking for vulnerabilities is valuable. Ideally, every school should establish a vulnerability assessment team that convenes at least once a year to conduct vulnerability checks, but at minimum, a one-time assessment provides baseline information. Look for diversity among team members, that is, individuals knowledgeable about different components of the school. With each vulnerability check and correction, the school's safety and response improves.
2. The vulnerability assessment should be conducted as quickly, discreetly, and calmly as possible. Students' sense of safety includes feeling safe when there's no disaster as well as when there is a disaster. It is important that students understand that the library is a safe place. If students become aware of preparation and planning for disasters as the librarian conducts a vulnerability assessment, it is critical that they understand the school is maintaining their safety by checking for vulnerabilities. Communication is critical in all phases of disaster and emergency planning.

Chapter 7

Drill, Drill, Drill

Until school administrators and law enforcement officials can commit the time, personnel and resources necessary for proper interagency planning and training, our after-action reports will continue to focus on lives lost and not lives saved.

—Matthews, 2014

Definitions of Terms

In this chapter some terms are simplified for brevity. "Emergency" and "emergency drill" are used as generic terms and refer to those plans that schools and districts design to provide exercises and activities in preparation for a real disaster or emergency. The term "drill," also generically used here, may refer to emergency drills as well as unofficial exercises to prepare for disasters.

Despite the ever-present academic pressures educators face in schools today, despite how time-consuming and disruptive the interruptions are, drills must be conducted. They must. State mandates, parental pressure, and the current state of external and internal threats to student safety create a need greater than ever before for frequent, ongoing, documented, and evaluated drills in every U.S. school.

While every district has requirements for conducting response drills, great variances exist in implementation. Whether drills are mandated from a centralized top-down authority or individual schools have flexible scheduling (time and frequency), types (which drills when), and postdrill evaluation that implements a revision process of the school's emergency plans and future drills, spending time within the school preparing for and carrying out drills will improve the effectiveness of any emergency response.

Prior to revising or updating existing emergency plans, consider surveying all staff regarding their participation preferences, training, and individual or unique skills. This simple act can provide a better plan and improved response during drills and in actual emergencies or disasters. For example, find out which teachers have CPR and first aid training. Perhaps someone in the school has outdoor survival, previous military, or amateur radio (HAM) training. Individuals may possess unique construction skills, counseling skills, or even organizational or management skills, all of which are important and can contribute to safety and provide the best use of the school's consolidated talent pool when it is needed most.

Once you have determined staff members' potential contributions to safety, preassessing each staff member's knowledge and understanding of the school's preparedness procedures (e.g., by implementing a school preparedness self-evaluation checklist) will help determine preparedness as well as areas of focus or concern.

Benefits of Emergency Drills

Emergency Drills validate existing emergency plans, programs, policies, roles and responsibilities, as well as the training curriculum, by:

- *Testing the capabilities of the team*
- *Providing a gap analysis to identify deficiencies and problems in the current program*
- *Familiarizing participants with plans and procedures through low-stress activities*
- *Preparing the team to coordinate with local, state and federal agencies*
- *Enhancing collaboration between school districts and public information officers*
- *Increasing the confidence of the team while strengthening its ability to respond effectively*

(Stockton Unified School District Emergency Response Crisis Management, 2009–2010)

STAFF PREPAREDNESS SELF-EVALUATION

Introduce a self-evaluation checklist prior to preparedness drills and operations to determine the level of preparedness of all staff members. This is most effectively conducted at the beginning of each school year. Be sure to include all specialists and support staff, office staff, substitutes, and classroom volunteers who frequent the building.

The following sections provide information about sample areas to include. Each building has its own unique needs and preparedness concerns depending on location, grade level, staff and student demographics, surrounding community, and other components of school enrollment and culture.

Sample Preassessment Self-Evaluation Preparedness Topics

Begin with the following four lines of questioning or areas to consider and create a school-specific self-evaluation preassessment. Doing this saves time, increases efficiency, and improves results for the emergency plan.

1. Do You Know Location Of:

- Your updated emergency procedure folder or notebook
- The class/student emergency roster, including individual needs
- An evacuation route wherever you are in the building
- Student and classroom emergency/comfort kits, or "go bags"
- The nearest fire extinguisher and fire alarm
- Procedures for replenishing emergency supplies

2. Do You Know:

- Your individual assignment or duties during drills/disasters
- Your emergency drill buddy class
- Your incident commander and alternate IC
- Your students' individual or unique health concerns
- Your students' custody, child care, or other family concerns
- Local/regional first responder information (e.g., police and fire)
- How to use a fire extinguisher
- How to lock the classroom or other room door in case of lockdown

3. Do Your Students Know:

- Evacuation rules
- Evacuation procedures
- Common drill vocabulary
- The location of emergency kits/go bags
- The assembly area
- The reunification location
- Cell phone use rules

4. Additional Self-Evaluation Questions:

- Have you told your administrator about any extenuating personal circumstances that would be relevant during an emergency?
- Are you prepared to address the unique needs of special needs students and ELL students during drills, lockdowns, or shelter-in-place situations?
- Have you spent time in class discussing safety concerns and expectations during drills?
- Is your personal contact information up to date (e.g., married, moved, divorced, emergency contact changes)?
- Have you planned sustained quiet activities for your class in case of shelter-in-place situations or lengthy lockdown?
- Are you familiar with state mandates related to emergency procedures?
- Have you addressed your at-home/personal emergency preparedness (see the discussion in Chapter 8 about personal documents)?
- Do you know what to do under certain unique circumstances, for example, an emergency when the principal or IC is out of the building?

Schools may need to compile and evaluate results from this preassessment self-evaluation to determine the school's next steps in preparing for drills. Alternatively, each teacher or other staff member may use results to determine individual class needs in preparedness. Tabletop discussion at faculty meetings may bring areas of concern to the front.

BEST PRACTICES FOR DRILLS

The first and best line of defense for school safety is a well-trained, highly alert school.

—Trump, 2011

Emergency drills have been conducted longer than anyone today has been teaching, but the review considerations regarding drills should be ongoing. Listed as follows are best practices worth examining and implementing where change is needed.

The Learning Curve

In schools where some drills are relatively new—lockdown or shelter-in-place, for example—librarians may wish to assume that ever-vigilant information specialist role and provide suggestions or resources for teachers and others who either want to have information to share with their students or need to read up on the importance of, effectiveness of, and current trends related to such drills. Understanding *why* drills are being conducted helps improve emergency drill performance.

In addition, the librarian may wish to provide information on the library website for parents and students about school safety drills and procedures. Include a few suggested links such as FEMA, the state or local emergency management office, and Red Cross for parent and child safety review.

PERSONAL STORY: Do as the Animals Do

Over the years when my wife was an elementary school teacher in Tacoma, Washington, I observed in the Pacific Northwest that many young children go to school not prepared to be outside for long periods of time in often cold, wet weather. I worried what would happen if there was a disaster, such as an earthquake, and they had to evacuate the building and it was deemed unsafe to return. Teachers and school administrators would figure something out after a while and many may already have disaster plans for such a contingency. But what do you do if you do not?

We had several teachers in our local districts come up with some ingenious ways to prepare. I especially liked one strategy that I call "Do as the animals do" that resembles behavior I had seen in Antarctica among the penguins. The teachers had each student add a large garbage bag to their individual emergency kits. A few holes in the right places make an expedient shelter. If they had to evacuate the school in bad weather and could not return, each student had a shelter that could protect them from the wet and cold. They then would huddle together on the playground and keep each other warm, and just like the penguins, take turns as to who would be on the outside of the group and who would be in the middle where it was the warmest. This could keep them reasonably comfortable until safe shelter was identified.

—Bill Lokey

Simplify

An emergency drill is one time administrators want all members of the school community to act in unison. Simplify as many of the procedures and actions for drills as possible to help ensure efficiency.

Simplifying student behavior expectations for evacuation will improve performance. For elementary schools in particular, create a simple jingle that students recognize as the expected behavior during drills. For example, "Voices off, eyes ahead, don't run, walk instead."

In addition to the district and schoolwide required emergency drills, abbreviated, simple classroom drills, conducted repetitively, will better prepare students for more complex drills simply because a few of the standard procedures during drills will become second nature. For example, if your school's duck-and-cover drill includes putting on helmets, you may use a simple helmet safety directive integrated with other activities or conduct quick helmet-only drills. In such cases students will soon be able to locate and put on helmets efficiently and then be ready for more complex directives during drills.

Standardization of Vocabulary

When seconds matter, all members of the school community must be able to respond instantly. One way of ensuring this is to standardize vocabulary and directives.

Everyone in the school must be able to interpret shorter directives quickly and identically. For example, the school's command for returning to class or seeking protection from open corridors prior to lockdown may be "Evacuate hallways," and students understand from this directive that they are to enter the nearest classroom for safety as quickly as possible. If the same directives are used repeatedly with each drill, students will know exactly what to do. In this example, they will understand that "Evacuate hallways" does not mean to go outside or go to another location; it means enter the nearest classroom for protection. The directives must always be interpreted in the exact same way.

Emergency terms must be part of students' working vocabulary. They must be used in a consistent manner during drills and practices. Exactly what is meant by key terms such as "lockdown" and "evacuate" may either need to be taught and reviewed frequently or simplified for younger students. Further, terms that have multiple meanings must be clarified, for example, what exactly is meant by the term "all clear"? Generic terms too must be clarified, for example, to what exactly is the "district command center" referring? Common vocabulary provides familiar emergency terms for all in-charge adults as well as students.

When vocabulary is standardized everyone involved is more likely to understand the status of an in-progress incident. When the vocabulary standardization involves the broader school community, first responders know what to expect upon entering the building. "The common vocabulary and protocols establish a greater predictability that persists through the duration of an incident. Parents can easily understand the practices and can reinforce protocol. Additionally, this protocol enables rapid response determination when [unforeseen events] occur. The protocol also allows for a more predictable series of actions as an event unfolds" (Keyes, March 24, 2014).

Methods of Communication

During drills, perhaps the most critical component with regard to all types of drills is the efficiency and effectiveness of all forms of communication. Make sure all lines of communication within and out of the school are clear. All in-charge adults must understand how to communicate with the main office and the incident commander.

Keep communication requirements for students simple. For example:

- No talking during initial directives.
- No talking during transition.
- No cell phone use (during drills).

Communication related to drills includes the local emergency response. If possible, involve local fire and police in the communication planning for drills as well as when drills are implemented.

Mix It Up

According to Ken Trump's school security blog, one of the most important lessons learned from recent school tragedies is the need to address as many conceivable disasters as possible, diversifying and making them increasingly challenging. "Conduct lockdown drills between class changes, during lunch periods, upon student arrival, at dismissal, during staff-only in-service days. Remove the building's leadership team and office staff to see how drills unfold without them" (Trump, 2014).

Involve the Village

A minimum of once a year, preferably at the beginning of the school year, a letter *specific to school safety* should be sent home with some response required or a directive for family involvement. Requiring a signed response ensures a greater likelihood that parents have read and understood the school's emergency drill procedures and safety precautions. One example is to include an informative letter when the school sends notification requiring families to provide items for emergency or comfort kits (go bags). This may also be a time when parents can be called upon to volunteer expertise and skills related to conducting drills and emergency procedures.

When requesting time for discussion of drills at faculty meetings and when suggesting increased time for drills within the school day, remind colleagues and administrators that the cost of holding emergency drills is minimal, and often it is free.

Most districts or schools send out notification of impending drills at some point. Consider involving families in emergency drills. For example, at a recent emergency preparedness conference, a principal shared her experience of requesting emergency parent pick-up. She practiced this "all school efficiency drill" several times during one school year, challenging families to beat their time for each drill, and she decreased her school's parent pick-up student release time by 200 percent.

Expect the Unexpected

While enactment of various scenarios can help older children learn certain safety procedures, there are some major emergencies or disasters for which no plans can be made and no predictable events unfold. When standard drills, regularly conducted, provide experience in various likely events, students are more apt to apply their emergency drill procedure protocol when a more unusual event transpires.

Special Needs Students

The postdrill evaluation for at least one drill each year should focus specifically on how well special needs students' unique needs were met. Schools may choose not to let staff members know about this predetermined evaluation focus. Following the drill, assessment of the effectiveness of the drill can focus on each student individually, the whole special education program within the school, or other special needs of one or more individuals not officially documented but known within the school.

Students with special needs are usually assigned an adult or a student buddy to assist them during an emergency. Drills offer an opportunity to identify who among the student population requires additional assistance. If a student buddy rather than an adult can assist in such emergencies, a drill is an excellent time to recognize or identify student leaders who may help share this responsibility.

Evaluate also how adults with special needs are cared for.

Sound the Alarm

One of the most important roles educators have during emergency preparedness drills is to acquaint students with the sound of the various drill warnings or the building's warning system sounds. Students need to understand the difference between a car backfiring or a siren outside the school and the school's fire alarms and other official warning systems within the school. Schools may choose to practice drill alarm responses in isolation to acquaint students with the school's official alarm systems and sounds.

Uniqueness of the Library

The library in each school is uniquely designed, located, and operated, a facility quite different from standard classrooms. The librarian may wish to sit at the table as school preparedness drills are designed or created to bring alternative and perhaps broader perspectives to planning.

FORMATS OF DRILLS AND TYPES OF RESPONSE

Critical to the success of drills is to mix criteria, focus, and complexity. The three types of drills most commonly used in schools today are tabletop exercises, functional drills, and full-scale drills. Each format provides a unique opportunity for preparedness.

Tabletop Exercises

Tabletop exercises, where key individuals sit at the table and discuss issues — particularly how the emergency response is going to be implemented — can be valuable for expediency and simplicity. Conducted without students, tabletop exercises can be informal and stress free while also being efficient and productive without movement and large numbers of individuals present. Tabletop exercises are primarily problem-solving exercises where various scenarios are suggested and solutions are discussed, and no resources or equipment are used. These exercises may be brief with minimal information (see **the appendix about faculty and staff tabletop drill scenarios),** or they may replicate extensive complexities of a major disaster (see **the appendix about scenarios in schools** and the **disaster priorities matrix).** These exercises are also good for refining components of more formal and complex and more inclusive drills. Tabletop exercises can be used as precursors to grander simulation drills and as a way to get all staff involved at different levels.

Functional or Simulation Drills

Functional drills simulate as closely as possible an actual disaster or school crisis, short of moving people from the school and bringing in local response personnel. These drills work to replicate disasters as realistically as possible with the immediate school population and on-site resources. The goal is to determine the capability of one or more emergency preparedness functions within the school in the context of a real event.

Full-Scale Drills

In full-scale emergency drills students experience the response to a disaster or crisis that is as similar as possible to an actual event. Often students are evacuated from the building and relocated to another site either through school district transportation or parent pick-up. Usually the first responders, police and fire, are also involved.

Response Procedures

In addition to the various drill formats, a comprehensive response includes varying response procedures. Different crises and disasters call for different types of response, including the following:

- Lockdown
- Evacuation
- Reverse evacuation
- Shelter in place
- Room clear
- Secure the building
- Drop, cover, hold
- Releasing students
- Relocating students

As schools plan their annual calendar of drills, each of these formats should be conducted at various times of the day, week, and school year.

TYPES OF DRILLS

Fire Drills

The schools in our nation have a long history of sound fire prevention and protection. Fifty years ago 90 percent of emergency responses by local fire departments in large cities were fires. Today 80 percent of emergency responses in those same large urban settings are medical and accident emergencies. Much has been done over the past 50 years to decrease the likelihood of destruction by fire. Even the simplest of technologies have greatly increased fire safety; fire alarms alone, for example, have saved thousands of lives.Despite technological advances related to fire safety, in some regions limited resources have led to fire safety negligence. Further, increasing arson coupled with the aging infrastructure of many schools has created a tinderbox of potential fire hazards. The open-space schools constructed in the 1960s, especially those whose convoluted student exits require multiple classes to exit through a single door, provide a greater fire hazard, particularly as they age.

> Some principals who are fed up with false fire alarms turn the system off, some chain doors shut because they are more worried about dangers outside the school than possible fires, and some chain gym doors to ensure that no one gets by the ticket taker at athletic events. Many schools try to skimp on the ten fire drills required by the National Fire Protection Association life-safety code, and many try to sidestep rules limiting the amount of paper that can be tacked to walls. (Jones, 1996)

Still, all school districts are required by law to conduct fire drills. For students and staff to fully understand the importance of fire drills as they relate to student safety, librarians might request a schedule of when fire drills will be conducted and then explain to administrators that fire drills can be made more meaningful if important fire safety content is taught in the library prior to the drills. Use emergency drills and safety content when teaching the information skills curriculum. Tell administrators when fire safety information curriculum will be taught and ask that a fire drill be conducted following this curriculum content.

Fire Drill Best Practices

Fire drills today are often considered of secondary importance behind drills for potential school violence or natural disasters. Yet, the likelihood of fire is far greater than an act of school violence or a major natural disaster. Librarians who have committed to increasing student and staff awareness of safety issues may wish to remind everyone of these basic fire drill best practices:

- *No one in the school is exempt from the drill, even off-duty staff.*
- *A calm and orderly exit from the building is more important than the speed with which all students and staff exit the school.*
- *The needs of special needs students should be clearly identified; following each drill, determine whether these needs have been addressed.*
- *The fire drill alarm's tone or sound must be the same as that for a real fire, and if possible, it should be a unique fire drill–only alarm sound.*
- *Fire drill evaluation: All adults working in the building should be informed of each fire drill's successes and challenges. If necessary, a repeat fire drill focusing on identified challenges may be in order.*
- *To better simulate situations that might occur during a real fire, schedule fire drills during recess, lunch, and even before and after school as well as mid-day.*

Flood Drill

Flood and fire are the two disasters that schools are most likely to face. Except for weather-related flooding such as flash floods, the immediate danger from floods is not to human lives but to the facility. Flooding, which can be either a human-caused disaster such as broken water pipes or weather related, ironically, is sometimes due to a fire. The wide range of causes of and dangers from floods at school may be motivation enough for adding flood drills, particularly for schools located near bodies of water with flood potential.

Like fire drills, full-scale flood drills cannot be implemented, but simulating a real flood as closely as possible provides students with skills they can use outside of school as well. Students need to know and be able to respond quickly to flood warning signals that are used in regions with frequent flooding. Incorporate these flood warning signals in simulation flood drills.

Post These Floodwater Danger Facts

- *Depending upon conditions, a person can drown in a few inches of moving water.*
- *Six inches of moving floodwater can sweep you off your feet.*
- *Stay away from power lines and electrical wires.*
- *Floodwater can hide depth changes—be aware of this when walking.*
- *Floodwater is more dangerous at night because of poor visibility.*
- *The depth of moving floodwater is difficult to determine.*

- *Use a stick or pole to "feel" your way forward in moving water.*
- *Drowning in floodwater is not the only danger: remember bacteria and pollutants, snakes and insects.*

It is imperative to work closely with local emergency response agencies to plan and execute both fire and flood drills.

Hostage Situation

Two scenarios related to staff may arise during hostage situations. Staff may be taken hostage and therefore involved in all parts of the crisis, or they may be aware of a hostage situation in the building. Hostage situations, like many intruder situations that are frightening for students, should first be addressed as staff-only tabletop exercises (see the appendix information about faculty and staff tabletop exercises). Once all staff members are trained in best practices, drills that include students may or may not be appropriate.

Response in a Student Hostage Situation

If you are the hostage:

- *Don't try to escape.*
- *Be very alert but move slowly.*
- *Try to remain calm and not cry.*
- *If you're the teacher with students present, tell them to do what the hostage taker instructs.*
- *Don't talk, but if you have to, speak calmly and respectfully.*
- *Tell your students to do likewise; ask them to follow your example.*
- *Take note of all details, including the hostage taker's change in tone.*

If a hostage is in the building:

- *Report if possible but do not intervene.*
- *Work to calm student fears.*
- *Note as many details as possible.*
- *Remember the person's physical features.*

Secondary School Student Stabbing

> *It is imperative for each of us to consider how we would survive such a violent scenario [such as a school stabbing] and to incorporate our survivor's mindset into our plans and subsequent actions. There are no guarantees in life. . . . However, there are preparations one can take in order to increase the odds of survival. The first step is to accept that a shooter could show up at your workplace. Once you accept that possibility you can plan for a reaction that may save your life.*
>
> —Kautzman, 2011

Because most schools have a weapons-free policy, it is unlikely you will use a replica of any weapon during drills. However, in secondary schools, drills need to address internal acts of violence. Implementing a stabbing rather than shooting incident in a drill may be less objectionable for some members of the school community and eliminate the need for a replica of a gun in the drill. The school stabbing response scenario in this book's appendix provides both students and staff with examples of drill experience.

The April 6, 2014, Franklin Regional High School (Pennsylvania) stabbing in which 22 student were injured is an unfortunate reminder that not only active shooter situations can lead to great numbers of victims; other weapons can be deadly as well. National school safety consultant Chuck Hibbert's response to this event reflects the importance of drills: "It does point out the need for training so this type of action does not occur again" (Hibbert, 2014).

Bomb Threat

Because bomb threats are most likely to be called in and therefore handled by the main office personnel, most schools do not include such situations in drills (see information in the appendix about faculty and staff tabletop exercises). Bomb threats lend themselves well to tabletop exercises and can be conducted quickly, often in a matter of minutes. Consider conducting tabletop bomb threat exercises with students working in the main office as well as office staff.

The feature about responding to telephone bomb threats provides best practices for responding to a caller.

Responding to Telephone Bomb Threats

While it is highly unlikely that a bomb threat will be called in on the school library phone line, threats happen across school programs and departments. The librarian may wish to share the following information with office personnel or have it on file for emergencies.

Appropriate verbal response/questions to bomb threat caller are as follows:

- *Where is the bomb located?*
- *When will it explode?*
- *What kind of bomb is it?*
- *What does it look like?*
- *What will cause it to detonate?*
- *Did you or someone else place bomb in our school? Why?*

The most important initial consideration is to remain calm, to listen carefully to every detail, and to keep the caller on the line as long as possible. Listen carefully to what is said and try to determine if the voice is at all familiar. The receiver of the call may be asked later to identify characteristics of the voice such as tone, speed, or distinguishing features such as an accent, lisp, or hysteria. Listen also for background noises. Under no circumstances should the receiver hang up the phone.

Postemergency Drill Evaluation

Although emergency drills promote understanding and growth, teams benefit most when post-drill evaluations are conducted immediately after the exercise. All stakeholders, from the school district level to the emergency response agency level, should participate in both the drills and exercises and postdrill event evaluations. The evaluation is most effective when it:

- Immediately follows the drill
- Includes everyone
- Engages the director of the drill
- Has an assigned recorder
- Includes a discussion of positive and negative outcomes
- Allows time for participants to discuss their observations
- Assesses whether the exercise's goals and objectives were achieved
- Creates an after-action report with steps for improvement.

Postdrill Evaluation Areas of Concern

Following each drill, evaluating the effectiveness of the drill will help improve performance for subsequent drills and provide a safer environment. One broad approach is to analyze the overall areas of concern. These areas may include the following.

What crisis management functions are most in need of rehearsal?

- *Evacuation*
- *Lockdown*
- *Communication*
- *Fire drills*
- *Violence management*
- *Announcements*
- *Coordination and control*
- *Weather drills*
- *Injury assistance*
- *Family notification*
- *Resource management*
- *Personnel management*
- *External agency interaction*

Postdrill People Locator and Summative Evaluation Forms

Use people locator forms to report any missing students or staff to the incident commander or crisis team members. The Risk RED School Disaster Response Drill Model and Templates created for the Great California Shakeout are available online and provide many excellent templates for emergency drill preparation, including a people locator form and post-drill evaluation (Petal & Green, 2009).

SUMMARY

> *I believe that it is our job, as a society, to protect these virtues in our young people. I want them to be worried about math tests and track meets and about science fairs and student council elections—the kind of normal school stuff that builds character. But our epidemic of gun violence is creating a culture of fear in our schools, where students are anxious about safety and intruders. These are worries no student should have.*
>
> —Lyle, 2013

While drills are one of the most important ways to prepare for emergencies, the exact details of any emergency situation cannot be predicted. Therefore, the better prepared a school is for anticipated dangers, the more likely the unexpected can be dealt with via the application of keen judgment.

An emergency preparedness plan is only as strong as its weakest link. Think of various drills as ways to find and correct those weaknesses. You may discover which staff members most easily fall apart during emergencies, or you could find exit doors that are blocked or claustrophobic students.

Educators' common sense and calmness are the most important emotional factors in successfully dealing with emergencies. These factors are not present by accident. Proper planning better prepares the school's students and adults to understand what to do in an emergency. The goal for conducting successful drills is that adults in charge will respond with levelheaded decision making so that if they must respond to an actual emergency, expertise drawn from experience will take over.

Chapter 8

Emergency Planning, Emergency Kits

Our schools have come a long way since [the school shooting at] Columbine, but Sandy Hook is a stark reminder of why we must never stop improving safety and security for our educational institutions in even the safest of communities.

—Dorn, 2013

After a disaster strikes, no one ever claims to have been too prepared. When events transpire, no matter how prepared schools think they may be, there's always something more that could have been added to the emergency planning.

Today one would be hard pressed to find a school in which students are not prepared with basics such as a fire safety curriculum, lockdown and earthquake drills, and emergency kits. Further, agencies such as FEMA and the Red Cross provide excellent online resources, which are updated frequently, for standard emergency preparedness for students. On their website the Red Cross provides detailed suggestions for the following:

- What emergency supplies to store, how many to have, and how to store them
- Classroom, individual student, and all-school emergency kit supplies
- Suggested building tools for emergencies
- First aid, medical, and sanitation suggested emergency supplies
- Budgetary considerations

AASL also provides several helpful standard checklists for emergency preparedness and responses specific to school libraries at http://www.ala.org/aasl/awards/beyond-words/site

Most of the ideas and suggestions in this chapter do not replicate the basic resources that schools already have; rather, they provide new ideas and suggestions, most of which are specific to libraries. Included also are preparedness processes and resources that are sometimes overlooked such as details regarding reunification plans. And finally, small actions that can have significant impact, such as hand cleaning and wearing helmets, are presented.

Ideas and resources here are not only for student preparedness, but also include suggestions that educators in charge should have in place in their personal lives. When disaster strikes, adults will be better suited to assist students in a calm and professional manner if they know their own personal lives have some security and that necessary precautions have been made.

PERSONAL STORY: Disaster Picnic

It has been an established best practice in schools, especially in the younger grades, for students to put together a personal emergency kit during the first few days of school. This kit includes emergency contact information, some nonperishable snacks, a small bottle of water, a garbage bag, a small toy or book, and other essentials personalized by student, class, and school community. Preparing this kit helps children learn about and understand preparedness and if needed, "buys some time" and provides some initial comfort in a disaster.

One teacher we knew added an extra item to her preparedness agenda. At the end of each school year, she would have a Disaster Picnic where the children would open their kits, enjoy the snacks, again talk about preparedness, and celebrate the fact that their school year had been a safe one. This reinforced the preparedness learnings she instilled in the pupils that school year. It also led many of the students to ask their teacher when school opened in the fall, "When are we going to prepare our disaster kits?"

—*Bill Lokey*

SCHOOL EMERGENCY KITS

Review the information in this chapter's features that addresses specific considerations related to various components of the school emergency kit. Then check to make sure your emergency kit is up to date with current standards. Certain universal "must haves" (content) and preparation processes (actions) summarize emergency kit preparedness and readiness.

When preparing for both school and community-wide emergencies, librarians must also consider their own in-house needs (see "The Librarian's Emergency Kit") as well as those in their personal lives such as personal documents (See "Personal Documents"). When disaster strikes, the well-prepared librarian can model emergency preparedness best practices and serve as a voice of reason and calm if her personal effects and life are in order.

Just as individual student emergency kits can be made more appropriate and valuable by individualizing a few items, so too for the school's administrator and teacher emergency kits. Personalized kits for administrators, teachers, and librarians should address the particular needs of the individual school as well, for example, towns or schools in tornado zones can include maps showing locations of nearby storm shelters.

Expiration Dates

Most schools prepare emergency kits at the beginning of the school year, and some eat the contents at an end-of-year picnic or donate items to a nearby food bank. During the school year, at least once, the contents of everyone's emergency kit should be examined. This may be a role the librarian wishes to volunteer for, holding a spontaneous emergency kit check by going class to class to ensure everyone has a kit. This task should also include checking expiration dates as appropriate. If items are not replaced annually, expiration dates should be checked regularly.

Universal Content: The Student Emergency Kit Must Haves

- *Water*
- *Ready-to-eat snacks or other food*
- *Prescription medications, if needed*
- *Family documents, photos, or memorabilia (something personal and familiar*)*
- *Hand sanitizer packets*
- *Personal identification and student information*
- *Engaging activities: games, cards, books (these may be for the whole class rather than individually compiled)*

**Do not include original irreplaceable items. In case of a fire, flood, or other destructive force, lost items of extreme importance can invoke additional trauma.*

Considerations for the Well-Prepared Student Emergency Kit

- *Kits' content items are well labeled (student's name and any important identifying information such as allergies)*
- *Kits include written instructions for items contained within*
- *Kits are light and easy to grab and go*
- *Kits are portable*
- *Nonperishables are excluded*
- *If possible, kits are personalized*

Water: The Secret of the Universe

The most critical item in emergency preparedness is water. Consider the following information related to accumulating, storing, and maintaining water:

- *Store one to two gallons per person, per day (FEMA recommends a two-week supply but a three-day supply is more realistic in schools)*
- *In hot regions, more water may be necessary*
- *Store in the coolest place possible*
- *Replace or check the expiration date every six months*
- *Store in presealed, food-grade, immaculate water-storage containers*
- *Do not reuse or refill containers*
- *Store water purification tablets*
- *During shelter in place for 24 or more hours, encourage children to drink water*
- *Do not ration water at the risk of dehydration*

If it is known immediately that water supplies are low, separate drinking water from other water sources (i.e., cleaning, cooking, etc.).

FEMA provides a detailed guide to water in emergencies at http://www.fema.gov/pdf/library/f&web.pdf

Items of Comfort for Student Emergency Kits

- *Family photo of happy times*
- *Small, named stuffed animal*
- *Representative toy such as a single Lego*
- *Individual note from parents*
- *Inspirational quotes and poems*
- *Personal religious or faith memorabilia (rosary beads, Bible, yarmulke)*
- *Pictures of important events in students' life with friends and/or family*

The Librarian's Emergency Kit

In addition to the standard items included in an emergency kit, some of the following may be appropriate for the librarian's emergency kit. Make selections according to personal need:

- *Reading glasses (why not include two pairs—another adult in the building may have forgotten reading glasses in his or her kit) or contact lens care items*
- *Extra hard-soled shoes (for walking through debris or water)*

- *Personal medication*
- *$20–$50 in cash in small bills*
- *Flashlight with extra batteries (include extra batteries for reading time)*
- *Reading material for students, teachers, and yourself*
- *Personal contact information: spouse or partner, children, pet's veterinarian*
- *A few self-standing candles and a book of matches (not for use in hurricane or other wind-related storms)*
- *Hygiene items for you and your students: travel/sample size tissues, towelettes, hand sanitizer, feminine hygiene products*
- *For secondary school librarians: sanitary napkins and tampons for students*
- *PDF of vital documents (see the information in this chapter about personal documents) on a thumb drive*

The Librarian's Emergency Preparedness Thumb Drive

It is amazing what one small thumb drive can hold. The comprehensiveness and compactness with which documentation can be stored on this small technological wonder makes it invaluable. Besides PDFs of personal documents (see "Personal Documents"), consider storing the following on a single thumb drive:

- *Prior to the first checkout of the year, take photographs—or, better yet, a short video clip—of all library resources. Walk around and videotape close-ups, close enough to read titles should secondary documentation be necessary.*
- *Take several still photographs of suggested book titles for reading during shelter-in-place situations. During a crisis, it may be difficult to remember or find the list of titles previously gathered for such an incident.*
- *Take a snapshot or have a list of critical phone numbers and contacts.*

Put these items on a thumb drive and attach to a lanyard to wear around your neck. Most of the suggested items are simply a backup to preparation that is already in place. However, having all of these on one hands-free, wearable, transportable, easy grab-and-go thumb drive may prove to be a most valuable organizational tool. This same documentation can also be stored on a personal cell phone.

Personal Documents

Floods, tornadoes, and other storms can cause complete destruction. Keep valuable documents at home or at high ground in a waterproof, fireproof, locked box. Having backup documentation can help educators recover their personal lives and thus be better prepared to help their students. Share the following list of important documents and other items that all adults in the school building should have someplace in safekeeping. If the library has a well-secured location to lock these up, it is appropriate to keep one set at school and one at home. Otherwise, second copies should be held at another secure site far from the original.

Vital Documents

- *Passport and copy of social security card and driver's license*
- *Wills and custody documents*
- *Copyrights*
- *Marriage, divorce, and death certificates*
- *Birth certificates, adoption papers, and citizenship papers*
- *Phone numbers of all important contacts, personal and businesses*

Financial Records

- *Insurance policy records and contact information*
- *Financial records, including savings and checking account numbers*
- *Credit card account information*

Home Records

- *Home inventory, preferably videotaped*
- *Home, house, and property records*
- *Important receipts, including automobile (and title)*
- *Health and medical records, including physician contact information*
- *Pet information*

Professional Documents

- *Educational documents*
- *Teaching license and/or certification*
- *Transcripts*

Weather- and Region-Specific Emergency Kit Items

In our mobile society individuals change residence more frequently than in past centuries. People often move to be near where jobs are, and librarians sometimes move to unfamiliar regions to be near their new jobs. The following list provides several region-specific emergency preparedness kit items librarians may wish to add to their emergency kits.

In regions preparing for cold shelter-in-place situations:

- *Emergency poncho or jacket*
- *Hand and toe warmers*
- *Warm socks*
- *Whistle*

In regions preparing for heat or drought shelter-in-place situations:

- *Extra water*
- *Dust or face masks*
- *Eye drops*

In regions preparing for catastrophic storms (hurricane, tornado):

- *Siren, whistle*
- *Helmet or hard hat*

In flood regions:

- *Rubber boots*
- *Waterproof gloves*
- *Water purifying tablets such as chlorine tablets*

In earthquake regions:

- *Helmet or hard hat*

Customized Emergency Kits for Kids: Student Voice, Student Choice

In order to get buy-in from students on the importance of creating an emergency kit for school and to help ensure they will do the same at home, suggest customizing kits by including items students want most. Involve students' opinions from the initial planning stage. In the preparation of emergency kits, FEMA recommends that students first write down on a piece of paper items they like to eat and activities they like to do.

They should also include ideas for a packable comfort item that will help them feel safe or provide comfort. If students are in the library during a lockdown, they may not have their classroom emergency kits with them. In place of a security or comfort item from their personal emergency kit, librarians can provide students with books to read for comfort.

> At the beginning of the school year, during library orientation, librarians should ask students to write down what book they would most like to have if they were stranded on a deserted island. Compile these lists and keep a copy in the library emergency lesson plan file or notebook. Having a list of these student-initiated titles can guide the librarian when reading to kids during a lockdown or if students are sheltering in place during a storm (see the discussion in Chapter 10 about books that heal).

Objects of Security

Many emergency preparedness websites and other publications that address children's needs during a crisis suggest having objects of security available to help reassure children who are afraid or help them overcome trauma. These items can also be used as diversions.

> In the elementary school library obvious objects of security include books. However, reading buddies—a bucket of stuffed animals—can double as objects of security when needed. Ask younger children to decide on a favorite from the bucket of stuffed animals and let them give their favorite a name, perhaps after their favorite literary character. Then each child can decide where his or her reading buddy or security buddy should reside in the library.

> In an elementary school library, particularly one in which the librarian provides planning time and therefore sees every class regularly, one bucket of reading buddies can serve multiple students from different classrooms depending on which classroom is in the library on any given occasion. That is, one stuffed lion might be Leo the Lion to Johnny in Ms. Jones's kindergarten class and the Cowardly Lion for Tara in Mr. Smith's second grade class. Many older students are unlikely to need a buddy until disaster strikes, but even middle schoolers will latch on to a soft familiar friend when faced with a crisis. Librarians reticent to display stuffed animals may simply keep a few in storage, on a back shelf, or in their emergency kit.

Comfort during Frightening Times

A critical consideration when children are in lockdown or are sheltering in place is children's understandings and beliefs about their own security, that is, their sense of safety. Research has shown that when students understand what is happening and their own concerns and fears are heard and responded to, they recover more quickly. "Schools are a core element of a community's fabric, serving as a source of continuity and structure, facilitating connections between community members, and providing stability during difficult times. Schools can serve a critical role in responding comprehensively to . . . children's needs" (Rossen & Hull, 2012).

Librarians can help provide a sense of comfort during critical times by providing a safe haven for students to express concerns related to school safety. As students share their fears, librarians can address their concerns. For example, when a major event happens and is televised nationally, allow students time to debrief and share their concerns; then clarify their understanding.

It is normal when an emergency strikes for those in charge to snap into action and give directives expeditiously and sometimes insensitively. While instructions must be delivered with a serious and what may be interpreted as an angry-sounding tone, for younger students this unfamiliar stern demeanor may cause greater distress and further fright. Attempting to use a calm voice when providing instructions about critical actions during a catastrophe may produce better results.

SUBSTITUTE LIBRARIAN LESSON PLANS AND EMERGENCY INFORMATION

Teacher-librarians always feel better about being absent from the library if substitute lesson plans are in order. Along with standard library procedures and plans, these plans should include emergency preparations. To ensure the substitute reviews the library's emergency material, librarians should clearly reference this material in the substitute's daily plans.

Substitutes, often unfamiliar with building routines, do not come to school expecting a crisis or other critical incident. Nor are they likely to have time to read through lengthy procedures and policies. Yet, on any given day small to catastrophic events may occur, and when emergency situations arise, actions happen rapidly. Therefore, emergency plan enclosures in the substitute file should be as brief as possible. Librarians may reference the building's various emergency procedures, or, better yet, provide a one-page summary of substitute emergency expectations for the library.

If the library has student helpers, make sure they are informed of safety procedures and the important role they can play in assisting a substitute. Or if the building has an ongoing safety committee that includes student safety officers, leave the names of these officers or student safety committee members to whom the substitute can make emergency policy or procedure inquiries, should the need arise.

Leave substitute guidelines also for standard emergency procedures such as the building exit plan, reunification plan, and the location of emergency preparedness equipment and kits.

In addition to standard emergency policy, librarians may wish to request that substitutes teach a safety curriculum lesson or read pertinent literature (see Chapter 10). These plans can also be used as emergency lesson plans when the regular librarian is suddenly absent without time to prepare plans. Lessons may be particularly useful in schools where librarians provide planning time for teachers and are expected to teach lessons daily. Since these lessons are designed in detail and are ready to use, they can provide substitutes and students with important safety content knowledge while relieving the absent librarian of the guilt of an unplanned day.

Locating Emergency Lesson Plan

Emergency situations often happened rapidly with no time for preparation. Substitutes are not likely to have time to sift through lengthy safety policies and procedures let alone try to locate emergency plans. Therefore, keep emergency lesson plans well marked and provide them as addendums in lesson plans for every absence. Include emergency preparedness content on a substitute lesson plan template. Two simple lesson plans for substitutes are hand washing and helmet safety.

Clean Hands

Most disaster preparedness publications do not include hand-washing exercises. But in fact, keeping hands clean is one of the most important ways students can keep from getting sick or spreading disease. Sometimes disasters come with potential health risks; some disasters are, in fact, health risks themselves such as a pandemic or a major flu outbreak. One of the most important factors in keeping students healthy and safe is reducing transmission of germs with thorough hand cleaning.

Hand cleaning is frequently taught as part of the primary grade health curriculum but is often forgotten as kids proceed through the grades. A quick lesson, regularly scheduled announcement reminders, or hand cleaning information posted in the library can help remind students of the importance of clean hands.

Remind students of these important hand-cleaning facts:

- For thorough cleaning let the water run until it warms.
- Make a lather and wash for at least 20 seconds.
- Try singing "Happy Birthday" twice through while washing.
- Rinse thoroughly.
- Dry with a blower or a clean paper towel.
- If soap and water are not available during a disaster, use an alcohol-based hand sanitizer.
- Alcohol-based cleaners work more quickly than soap and water but provide less coverage.

Always wash:

- After using the bathroom
- Before eating
- After handling animals or pets
- After handling garbage
- After blowing noses
- After coughing
- After sneezing
- Before treating a cut
- When caring for someone sick
- When hands look dirty!

Helmets for Safety

The nation has come a long way since a generation ago when most children rode their bicycles with no head protection. Helmet safety is one of the simplest safety standards educators can put into place. Bicycle helmets can also be used as protective gear during a natural disaster.

Greater safety measures are often put into place following a disaster. After Jennifer Bromley's son was nearly killed in the 2013 Moore, Oklahoma, tornado (he resourcefully used a library book to cover and protect his head during the crisis), she organized the Moore Helmet for Schools fundraiser to ensure every child in the 34,000-student school district has a helmet to wear during tornadoes (Breslin, 2104).

While the safest thing for students during a tornado is taking cover in tornado storm shelters, many injuries, particularly head trauma, occur from falling debris, protection from which helmets can provide. Emergency relief organizations are now recommending helmets as part of standard emergency preparedness gear.

PERSONAL STORY: No More Second Grade Assemblies

During the time I was the director of emergency management for Pierce County, Washington, we had members of my staff go to schools to give preparedness assemblies. They were very popular, but I soon realized that with 19 school districts and hundreds of schools in the county, this was very ineffective. We just did not have the resources to reach all of the children in a sustainable way.

We redirected our efforts to training the teachers and providing materials so that they could run the assemblies. We developed and presented train the trainer workshops for teachers and school staff and soon were reaching thousands more pupils with our preparedness message than we had been.

We worked with federal agencies such as FEMA, the National Weather Service, and the U.S. Geological Survey to get information and take advantage of their training programs to develop a cadre of trainers who could sustain our outreach efforts. We knew we were being successful when we received a letter from a second grade teacher who had her class do reports after the earthquake safety presentation (see "Letter from Second Grader").

—Bill Lokey

Role-Playing Safety and Emergency Situations

One of the best ways for students to train for emergency preparedness and to understand emergency procedures is to role-play scenarios related to the safety information they need to learn. While it is important for students to understand the gravity of emergency preparedness, librarians can make this an enjoyable *and* educational activity. Often role-playing, an activity so common in primary grades, is not used as an instructional method in upper grade or secondary classrooms. Librarians may be surprised, however, at how well students embrace highly interactive role-playing exercises. Further, when students become engaged with the learning activity, they are likely to learn material more deeply and adopt a preparedness disposition.

Below are some short, simple suggested role-play scenarios that librarians may use with students as lesson introductions or ice-breakers, as quick filler activities, or as part of a longer safety curriculum unit. Be sure to follow each activity with a discussion of the experience, including how students might apply the experience to their own lives. Remember that these are only sample role-play scenarios. After trying these, let students create and act out their own safety scenarios. Some may wish to use emergency planning or safety issues unique to the school's culture.

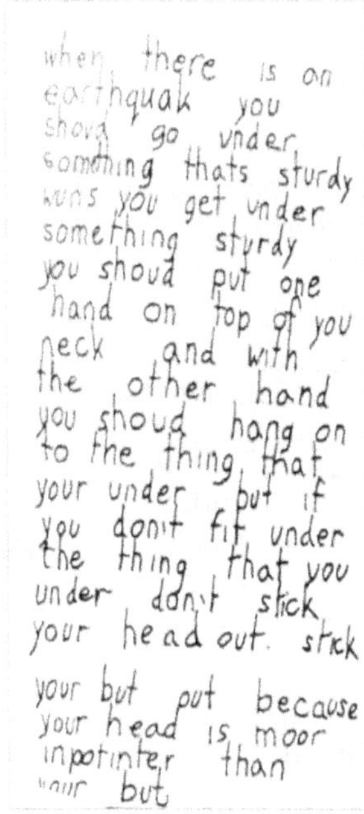
when there is an earthquak you shoud go under somthing thats sturdy wens you get under something sturdy you shoud put one hand on top of you neck and with the other hand you shoud hang on to the thing that your under but if you don't fit under the thing that you under don't stick your head out stick your but out because your head is moor inpotinter than your but

One of the letters the Pierce County Emergency Management Staff received from second grade pupils after their Earthquake Safety presentation at a local elementary school.

Scenario 1: How to Contact Emergency Services (grades 1–3)

Imagine your class is in lockdown and your teacher has suddenly been called into the hall because of some unknown emergency. Upon leaving she quickly says, "Don't anyone leave!" It is unlike her to not give assignments before leaving the room, and she has not returned after some time. After a time, students suspect something is amiss. On the teacher's desk is a telephone. What should you do?

(Let each student go to the telephone and practice a call to emergency services. For this exercise, particularly for first grade classes, the teacher may stand to the side and role-play the operator for emergency services.)

Assessing Success for Scenario 1

- Students know to call 911.
- Students describe exactly what happened.
- Students are able to answer questions from the emergency services personnel.

Scenario 2: A Student Is Hurt during Recess (grades 2–4)

You are out on the playground during recess, and the friend you are playing with is accidentally hit in the head by a ball from the softball game being played on the field. Your friend falls to the ground and does not get up. You look around and do not see a playground supervisor. What do you do?

(Partner students and after they have had time to discuss the scenario with their partner, let them alternate between playing the hit child and the friend. Follow up with a class discussion.)

Assessing Success for Scenario 2

- Students know not to move someone who has been hit in the head.
- Students are able to seek further help.
- Students stay or have someone stay with the victim.

Scenario 3: Approached by a Stranger when Walking Home from School (grades 3–6)

You and two friends are walking home from school. Suddenly a car pulls up very close to you. A stranger hops out of the car and walks toward you. He begins talking to you from a distance, asking for help with directions. He's walking quickly and with confidence as he talks.

(Assign groups of three to five students to first discuss the scenario and then decide how they will react. Then they can share their responses with their classmates.)

Assessing Success for Scenario 3

- Students know not to respond to the stranger.
- Students understand they must stay close together.
- If necessary, students know to run away rapidly and/or scream.

Scenario 4: Power Outage in the Classroom (grades 4–6)

Your class has a substitute teacher and suddenly, during a wind storm, the classroom loses electricity. The substitute is unfamiliar with other parts of the school building or building emergency procedures. What do you do?

(Make groups of three to four students each and ask students in each group to work together to prepare and act out their response. Responses will vary, and issues that may need to be addressed will come up.)

Assessing Success for Scenario 4

- Students understand the importance of their leadership role when a substitute needs their assistance and their knowledge of the school.
- Students know to contact adults in charge when in doubt.
- Additional understandings will vary depending on unique characteristics of the school and community.

Scenario 5: Your Front Door Is Standing Wide Open as You Arrive Home from School (grades 4–8)

Typically when you arrive home from school, someone (parent, older sibling, neighbor, sitter) is there waiting to greet you. But as you arrive home and approach your house one cold winter day, you notice your front door is uncharacteristically wide open. What do you do?

(Students may respond to this scenario in various ways. Students may be grouped and just discuss what to do, or they may role-play.)

Assessing Success for Scenario 5 Student responses will vary depending on personal and home situations; however, some key understandings include:

- Students know not to enter the house if they suspect foul play.
- Students will seek help from neighbors or emergency responders (911).

Scenario 6: Gang-Related Potential Attack (grades 5–7)

After school you are walking home alone and notice on the sidewalk ahead a group of older kids, not your friends, who have been identified as local gang members. They are watching you, murmuring comments to each other and laughing. As you approach, one of them shouts at you as he begins walking aggressively toward you, "Hey, we need some gas money!" You suspect he is planning to shake you down.

(Let students discuss actions they would take. The direction of the discussion will vary depending on school culture. However, this scenario may also provide the opportunity to discuss profiling and the dangers of judging by appearances. Expect varied responses.)

Assessing Success for Scenario 6

- Students know to leave an unsafe scene quickly.
- Students are able to recognize extortion.
- Students learn that informing school authorities of extortion is critical to a safe school.

Scenario 7: Personal Identity Theft (grades 7–9)

You come home from school, and your mother explains she has been contacted by legal authorities about your online account (e.g., Instagram or Facebook). She asks if you have applied for a credit card or other account. You realize you need to increase the protection of your online identity, change passwords, and check all other activity/accounts.

(Have students partner with someone they trust and discuss this scenario. They may share which online accounts they have and the processes they have used to protect their identity. Afterward let them share ideas they've discussed, being careful not to share their own personal safety information during the discussion.)

Assessing Success for Scenario 7

Children know how to check activity on their account.
Children know how to password protect their online account(s).
Children know their identity is a potential target for theft because:

- Identity thieves know children typically do not monitor their online identity.
- Children's personal identification numbers are typically unused or not associated with other active accounts.
- Children are often less suspicious than adults with online identity.

Scenario 8: Potential Active Shooter (grades 7+)

You are walking down the hall alone during class, and you suddenly hear a loud popping noise. You suspect it might be a gunshot. You hear several screams that further confirm your suspicions.

(Students need to recognize the gravity of this scenario. To ensure seriousness, it is not recommended that students role-play. Rather, students should discuss actions. The librarian may wish to assign discussion partners.)

Assessing Success for Scenario 8

- Students seek immediate protection.
- Students know that if they have a cell phone, they should first text 911, then text family, and then silence the phone.

Each of these role-playing activities can be performed in a brief time frame. Perhaps students who have completed library checkout may wish to prepare a scenario response to share with classmates or their teacher when he or she arrives to pick them up and return to class. Alternatively, students may design their own scene or safety role enactment.

REUNIFICATION PROCEDURES

While there is a smattering of science in [reunification planning] methods, there is certainly more art. . . . Crisis recovery starts with *the crisis, not after. Simply "winging it" when reunifying ignores not just the mental health demands which accompany a crisis, but the responsibility of the school and the district in maintaining the chain of custody for every student. No school is immune to stuff hitting the proverbial fan.*

—Keyes, February 20, 2014

A reunification plan is "an orderly and efficient process of reuniting children with their parents/guardians after an emergency/crisis which prohibits the normal dismissal process from occurring" (Rock View Elementary School, August 20, 2014). All schools have an emergency preparedness plan in place, but not all emergency plans carry through to the all-important point of reunification. If the reunification plan is well prepared, well rehearsed, well communicated, and followed, children who are not directly impacted by the natural disaster or crisis in the school may have minimal long-term negative impact from the event. Conversely, while a child may not suffer any personal tragedy related to an event, since separation from family is one of the greatest fears children have during a disaster, without contact with parents and expeditious reunification, even mild or false alarm incidents may have a chilling and frightening impact on some children. Further, without a clearly designed reunification plan in place, chaos happens very quickly and—even in the event of a false alarm—can create or magnify an event as parents and guardians with good intentions try to reunite with their children.

Typically, designing and implementing a reunification plan is a one-time event that, after it is put into place, may have to be updated only once or twice a year. One role that the librarian can assume is taking the lead in developing and maintaining the reunification plan.

Some emergency preparedness experts suggest creating a centrally located reunification plan that includes student records, including their emergency contact, custody, and medical information. While it is recommended that original reunification plan records be housed in the office, as with all emergency preparedness, a backup location is wise. The librarian may volunteer to maintain that second or backup location for student reunification information if the library has a secure facility available.

> Include local agencies and community members when designing and establishing reunification plans. Inherent in the planning when it involves the broader community is strengthening partnerships and communication with local agencies and first responders.

A subset of creating the reunification plan is language and vocabulary. As a member of the reunification design team, the librarian may wish to accept this role. Ensure there are no language barriers with students and families who do not speak, read, or write English as their first language. As with all school communication, translation of reunification procedures must be disseminated for families in all designated languages.

Further, ensure a shared vocabulary with all members of the school community, including younger children. Simplifying terms will help diminish students' fears, particularly the school's youngest children, special needs students, and ELL students. For example, "reunification" as described to younger children may simply be called "finding your family."

> When the reunification plan is put into place, there are numerous roles to be filled. The librarian who chooses not to hold an integral role in reunification planning and implementation may wish to volunteer for an incident or drill role during reunification implementation such as a runner, social media announcer, information provider, parent greeter, student or parent identification personnel, or crisis detractor. Or the library may be used to provide a "holding center" for students whose families the school is unable to locate.

Well-Designed Reunification Plans

- *Provide clear procedure for releasing students*
- *Include designate reunification sites*
- *Maintain updated student emergency information that includes medical or custody issues*
- *Ensure storage of student information in a secure location*
- *Provide clear notification plans with multiple communication methods, including but not limited to online communication plans*
- *Involve parent and student input throughout the planning stage*
- *Involve communication and planning with local outside agencies*
- *Include proven methods for planning, practicing, and achieving a successful reunification*
- *Will be updated as needed throughout the school year*
- *Clearly identify and communicate roles*

As with all plans: keep it as simple as possible for ease of implementation

Reunification Plan Design Considerations

- *Signage*
- *Parent check-in area requiring valid ID*
- *Student nametags*
- *Situational awareness communication plan*
- *Parking*

By having a defined process with signage, cards, branding, procedures and protocols, the school presents an organized, calm face to all involved. Fear or uncertainty often results from the unknown. By adopting, communicating and practicing a "known" procedure, the school removes some of that uncertainty. As a result, better decisions occur, leading to successful reunification. (Keyes, 2014)

Daily Release versus Controlled Release

Depending on the school's location, children arrive and depart in a variety of ways. There are times during an emergency that it is inappropriate to allow students to leave the school building in the usual manner of daily release. That is, there may be community precautions in place preventing the safe release of students or safe routes home.

Controlled release may be implemented because of inclement weather, a hazardous material incident, criminal activity or child predator information release, custodial issues, or other student safety issues. A controlled release is also usually implemented within the school day, when the release time is contrary to standard dismissal arrangements. With controlled release mass communication via phone or Internet may be required.

Some school districts devise alternative reunification plans for controlled release of students. This added precaution, while it may prevent students from going about their normal route home, may require additional contact information as well as neighbor, community, and extended family emergency pick-up plans.

As often happens following a tragic event, a community of school members may decide to help improve some components of emergency planning by redesign or new implementation to better ensure student safety. One such tragedy involved Emily Keyes of Bailey, Colorado (see the feature in this chapter). As a result of this tragic event, one of the most comprehensive reunification plans was developed. It can be found on the website, www.ILoveYouGuys.org.

Emily Keyes's Story

On September 27, 2006, a 53-year-old armed man took seven girls hostage inside their high school in Bailey, Colorado. After several hours of negotiations, the gunman shot and killed Emily Keyes before killing himself. When Emily's father initially heard of the hostage situation, he texted his daughter, who responded, "I love you guys." This last communication John-Michael Keyes had with his daughter is now the name of the foundation the Keyes family started.

Following this tragedy, members of the Keyes family collaborated nationwide with school members, community members, first responders, and mental health agencies to create one of the most comprehensive reunification plans, which can be found at the "I Love You Guys" Foundation website.

The mission of this organization speaks to the restorative power of a family committed to making positive change from a tragedy: "The I Love You Guys Foundation was created to restore and protect the joy of youth through educational programs and positive actions in collaboration with families, schools, communities, organizations, and government entities." (Keyes, 2014)

The "I Love You Guys" Foundation has expanded to include resources for organ and tissue donation, an antibullying campaign, information about the dangers of sexting, and other violence prevention. It has also established college scholarships for acts of kindness. (Keyes, 2014)

SUMMARY

For a school to be as prepared as possible for any inevitability, a culture of preparedness must be established. Ideas put forth in this chapter are meant not simply to broaden educators' perspective on ways to keep students safe but, more important, to empower students whenever possible to prepare *themselves* for emergencies. It is critical for students to know they are safe *in their school*, but they are not in school 24/7. Librarians can work to create a safe haven for children in the library, but they must also work to empower students to take charge of their own safety.

Ideas included here are also meant to show that providing opportunities for learning emergency preparedness and safe practices do not have to be formal lesson plans with assigned class work and more importantly should not be taught solely in isolation. An attitude of preparedness can be something as simple as reminding students of everyday safe practices like clean hands and wearing helmets when necessary, and providing opportunities for students to discuss everyday encounters.

From the librarian's perspective, preparedness means integrating safety curriculum into daily practice, proactively modeling behaviors to safeguard children, and creating respectful classrooms where students feel safe, heard, informed, and respected. It also means providing broadly inclusive avenues of response that incorporate families and community-wide participation. A culture of preparedness offers a physical environment—the library—where children feel protected and safe.

Columbine: The Road to Recovery Has Led to Success

What changed in the U.S. with regard to the Columbine High School shooting was that it was the first such event that was shown live on television. The nation watched stunned as students fled from the building, running for their lives.

—John McDonald, executive director, security and emergency management, Jeffco Public Schools

After the 1999 Columbine High School shooting in Jeffco Public Schools, a number of directors of safety came and went. They were hired to re-establish an academic setting for the district as well as increase security measures and lead recovery efforts. It was a near-impossible job. Scenes from the day of the shooting, given national attention, were repeatedly broadcast for months afterward, shining a far too bright spotlight on every action taken following the tragedy. The job of director of security for the Jeffco schools was enormous; responding to student, staff, and community needs along with overwhelming media attention made any return to normal life impossible.

Eventually—and fortunately—the district found a security director with law-enforcement and active shooter background who had a personal interest in restoring normalcy to Columbine in particular and the entire district as well. In addition to working on security measures for the 2002 Olympics and subsequent national political conventions, John McDonald's prior training included leading disaster response teams. Since coming into the position with the Jeffco schools, McDonald has worked closely with law-enforcement personnel and community members to develop one of the nation's most successful school safety and security plans. "In the school, teachers, librarians, coaches, are all first responders," McDonald reports. "Our [safety and emergency preparedness curriculum] begins in kindergarten and is taught and practiced in every classroom, every year. Drills are [diverse and more closely replicate] what may happen on any given day. Drills are not just for when students are in class. They need to know what to do if they are stuck in the hall or the gym, or other vulnerable place during an attack."

Safe2Tell, Safe2Text

Shortly after joining Jeffco School District's administrative team, McDonald recognized the need for early intervention and increased connectivity with students related to information about security and crisis prevention measures. He implemented a strategy for a texting component of an early warning student anonymous hotline called Safe2Tell, a Colorado program that was created as a direct result of the Columbine Commission findings. More than 32 states have expressed interest in this one-of-a-kind school safety hotline.

"The students know what is going on in the school long before the adults," McDonald stated.

The success of Safe2Tell is due in part to several important understandings. First, students know that their tips are completely anonymous. When students call in a suspicion—whether it is information about bullying or a potential suicide or other school threat—the information immediately goes to the school principal and local law enforcement. The expertise of these professionals then determines next moves and threat credibility. Prior to McDonald's arrival at Jeffco schools, the 86,000-student school district normally had an average of 39 student tips per year. Once Safe2Tell was implemented and Safe2Text was added (so that students could also text tips), tips increased to over 500 per school year. Currently the greatest number of these tips are suicide threats.

Safe2Tell has successfully prevented 28 planned school attacks just in Colorado in the 10 years since its inception. McDonald reported, "These were no small threats. Information contained in evidence retrieved included hit lists and arsenals of weapons. These serious threats contained documentation of potential victims, as well as details about the planned attacks."

In addition to the hotline, Jeffco surveys its students about their sense of personal safety at school. The annual survey, Make Your Voice Heard, gives every student the opportunity to express his or her sense of safety and any concerns they have.

Safety Leads to Academic Success

No one can convince McDonald that time spent on safety measures takes away from academics. In Jeffco, safety *is* academic. "At 84%, our school district has one of the highest graduation rates among large school districts in the nation. This is due in part to the fact that students in our schools feel safe." A sense of safety is a basic need of all children, both in school and out.

Working Together

"The culture of the country is different today. In Colorado we are averaging a school shooting every 2.8 years. Within a 35-mile radius, there have been five school shootings since the 1999 Columbine tragedy." Because of this, McDonald insists, safety standards must be a top priority for every school and every district. This also means that every member of a community should be involved in school safety measures. In Jeffco, families, law-enforcement personnel—particularly the Department of Security and Emergency Management and local law-enforcement officials—and the schools work closely together to ensure student safety.

Columbine Today

McDonald acknowledges that a culture shift happened in Columbine on that fateful day. "We have never served the same meal that we served on that day. We have kids whose parents were here during the shooting. And teachers teaching at Columbine who were students here on that day." But while 14 years later "Columbine" may still be associated across the nation with that tragic school shootings, and although residents have not forgotten, it is a very different community. Students and staff take pride in their school. Columbine is a highly successful high school with excellent academic and graduation rates. And it is probably one of the safest communities in the nation.

A special thanks to John McDonald, current Executive Director of Security and Emergency Management at Jeffco Public Schools for his time and expertise in this report.

Chapter 9

Recovery

We are Sandy Hook. We choose love.

—Signs displayed throughout the Sandy Hook community following the 2012 school shooting

When a major disaster befalls a school community, educators quickly discover how complex the recovery process is. Children respond in so many different ways that trying to determine whose needs are greatest becomes an overwhelming task. Is the student who has an outward display of anger or the student who has withdrawn and become uncommunicative most in need? Should educators attend to those students whose parents advocate strongly for assistance and recovery services or seek out students whose responses to disaster are more subtle?

Children are often more resilient than adults. True, after a traumatic event some may continue to have difficulties and others may be impacted long after the event has occurred, but *most* children will recover from even the most tragic or horrifying events. Children's reactions, particularly during extreme situations, are greatly influenced by the adults around them; they rely on the strength of the important adults in their world. Therefore, to start with, although it may not be easy, one role the librarian can assume following a disaster is maintaining a consistent, honest, communicative, calming demeanor. Ensure that all children know at least one place they will find a safe haven: the school library.

When an entire school or community has experienced a major crisis, all functioning systems in society will be impacted. Undeniably, the school culture will be changed forever. Realization and acceptance of this fact is the first step toward recovery.

The Most Vulnerable Children

Because immediate first-responder recovery efforts must focus on the primary victims of a disaster, secondary victims' reactions can often be overlooked. Following a disaster children who were secondarily impacted and are now more likely than others to be at risk for developing long-term disaster response issues include:

- *Children who witnessed the event*
- *Children who had some firsthand exposure to the event*
- *Children with prior significant family or personal stressors*
- *Children with previous loss or trauma experience*
- *Children whose adult family members have difficulty dealing with the event or inadequate parent or guardian support*
- *Children who thought they might die during the disaster*
- *Children with pre-existing mental health issues*
- *Children who had to relocate following the disaster*
- *And, of course, those whose family members were personally impacted, including both children whose family members were victims of the event as well as those whose family members were personally responsible for events that transpired*

Children's Varied Reactions: Recognizing Their Response

> *Well informed teachers and school personnel can be a source of support for students, but unintentionally they may, if not well informed, be a source of stress. For example, not understanding reactions may lead to misinterpretation of behavioral reactions to loss as misbehavior and disrespect for others.*
>
> —American Association of Pediatrics, 2014

In response to a major disaster or other school crisis, local or state social and health services will likely provide professional development training and assistance for all educators within the school. Those who are charged with recovery efforts related to social, health, and psychological post-event services can offer individual suggestions to adults in the school community. By its very nature, a major event draws the school community together as each individual adult assists with these services to the capacity of their role and training. Yet, according to *Children and Disasters: Preparedness, Response and Recovery,* which contains a report to President Obama and Congress by the National Commission on Children and Disasters

> As a result of limited access to formal mental health services and treatment following a disaster, communities depend upon persons who are not mental health professionals but who routinely interact with children—such as teachers and staff, first responders, healthcare professionals, early education providers, and members of the faith-based community—to provide basic support services and interventions. As indicated in the Commission's Interim Report, these individuals must have basic knowledge of how to recognize signs of distress, assist children in adjusting and coping, and identify children who require more advanced care. (Walker, 2012)

As the school's information specialist, the librarian can help connect adults to outside resources for learning about and recognizing children's responses to traumatic events and thus reach a wide number of school community members who are engaged in recovery efforts. This information can be accessed before a disaster occurs from local and/or state emergency management agencies so that the school librarian is prepared to assist adults with these services. The worst time to try to find out about such services is in the aftermath of a disaster when emotions are high, the agency's forces are stretched, and adults are focused on many immediate needs.

How children react to stress following a traumatic event depends greatly on their developmental level. For example, young children such as preschoolers may show easily recognizable outward displays such as tantrums, unexplained or inappropriately severe outbursts, crying, fear of the dark, and clinginess. They may also revert to earlier childhood behaviors such as thumb-sucking or bedwetting.

Younger Children's Observable Responses after a Disaster

The following, drawn from multiple studies on the impact of disaster on young children, are observable responses:

- *May fear separation from family, particularly parents or guardians*
- *May need reassurance regarding images seen on media*
- *May personalize news stories and believe the events are related to their own lives*
- *May have difficulty keeping facts of the event in perspective*
- *May suffer guilt believing such events were caused by their own behavior*
- *Very young children may not be able to separate reality from fantasy or fiction; older elementary kids who may understand the differences under normal circumstances may have difficulty separating fantasy from reality during particularly stressful or fearful events; some may have difficulty distinguishing news footage from scenes in a scary movie or television show, for example*

- *May not understand the concept of "rebroadcasting" and may assume the event is happening again*
- *May be conflicted about results of the event with regard to punishment of the perpetrator, or they may focus on the injustice to victims of the event or disaster*
- *May have difficulty with the concept of geographic proximity, that is, assuming an event or disaster is nearby; for example, preschool and even primary age students may see footage of war and assume or fear closer proximity*

As elementary age children develop greater abilities to communicate clearly and control emotions, the manifestations of their stress may take either a verbal or physical form. Examples of verbal response at this age include denial ("It didn't happen"), personal blame ("It's my fault"), repeated unanswerable questions ("Why?"), fear for their own personal safety ("Will I be safe/harmed?"), and describing personal physical ailments related to stress, the most common of which is stomachaches. Students may also demonstrate moodiness, withdrawal, or crying jags. Physical manifestations of trauma may include aggression or anger and may in fact lead to fighting and even bullying. Educators may find trauma impacts students' academic work, concentration, and attendance. An indication of a positive response to a critical event may be discovering children engaging in productive activities or seeking to improve conditions for those impacted or others around them who are in need.

Teenagers' response to traumatic experiences may mirror many of the symptoms mentioned earlier in this discussion. Interestingly, teens, who are developmentally at a stage where friends and classmates are often the most important people in their lives, may actually distance themselves from their friends in response to trauma. Teens are also more likely to suffer depression, anxiety, mood swings, and potentially even suicidal thoughts. Even for teens who have no history of risky behaviors, some may turn to substance abuse as a coping mechanism. They may also participate in other risky and even illegal activities. Teens who ruminate over dramatic events or become highly self-preoccupied need attention and probably counseling to help them move beyond destructive thoughts and consuming self-absorption. At school, as with younger children, teens may suffer a decline in academic performance, poor attendance, and difficulty concentrating.

Universal Traumatic Stress Reactions in Children

According to About our Kids *an online recovery resource prepared by faculty and staff at the New York University Child Study Center, after a disaster or major crisis, educators should look for the following symptoms as common traumatic stress reactions in children of any age:*

- *Nightmares, particularly about the event, and other sleep problems*
- *Agitation*
- *Nervousness*
- *Difficulty concentrating*
- *Irritability*
- *Change in appetite*
- *Disorganized behavior*
- *Withdrawal from friends and social interaction*
- *"Memory loss for information about/blocking out negative details of the event.*
- *'Flat' emotions or 'numbing' of emotional expression*
- *Re-experiencing*
- *Repetitive play or acting out of traumatic related events*
- *Acting as if the event is occurring again" (McClough, 2006)*

Survivor Guilt

It's hard to imagine what might now or in the future go through the mind of the six-year-old who was the one and only child who survived the second grade classroom shooting at Sandy Hook Elementary. Certainly her grief and possibly her guilt will follow her the rest of her life. Depending on results of a disaster or school crisis, some children suffer survivor guilt. "Survivor guilt prompts thoughts about how the loss of a loved one, co-worker, or even a stranger might have been averted. These are normal reactions but tend to involve unrealistic ideas ('Why have I survived when so many did not?') that can result in self-blame and difficulty coping" (Farrell and Crimmins, 2001). Much can be done through counseling with mental health professionals to help an individual recognize and alleviate survival guilt. With natural disasters as well, children may have difficulty understanding why their lives have been less impacted than those of their friends. "Some children express guilt that they still had a house to live in while their friend next door did not" (Brock, Lazurus, & Jimerson, 2002).

Response Variables

Following a disaster or other significantly traumatic event, children may respond in a variety of ways depending on:

- *Age and gender (girls are more likely to suffer from PTSD than boys [Giannopoulou et al., 2006])*
- *Innate ability to deal and level of resilience*
- *Understanding of the event*
- *Level and type of support: family, community, and school*
- *Personal communication skills*
- *Previous exposure to trauma or loss*
- *Mental stability prior to the event*

Librarians Helping Students Cope and Recover

Librarians are not trained as school counselors, nor are they likely to have trauma response training from the field of psychology. But while they should not be expected to provide such counseling services, there's much librarians can do, especially from the safe haven of the library, to help students move toward recovery following a traumatic event. First, much research on children's response to trauma indicates that students who are surrounded by calm, consistent, and emotionally stable adults are better able to assimilate recovery efforts themselves. Simply providing a sense of **stability** can be a great service to children recovering from a traumatic event. When adults around students are grieving, anxious, or otherwise unstable, children absorb and often adopt these responses as the norm.

Librarians can **provide data** and statistics to reassure students of their safety and security. Following critical events, particularly a natural disaster, there is a tendency to assume such events happen with much greater frequency than they actually do. Remind students—even those in regions with a high probability of tornadoes and hurricanes, for example—what the actual risks are. This more realistic outlook on the real dangers from disaster—natural or otherwise—will help assuage children's fears.

Another opportunity for the librarian during recovery is simply being available to individuals as a **listener.** Children may need a sounding board for their stress, fears, and anxieties. Letting children voice their concerns in an appropriate setting may help move them toward recovery. Recovery experts agree that it is critically important for a key adult to be both physically and emotionally available to continually students after these events. Often, however, when the entire school has been affected by a traumatic event, there are simply not enough adults available for the number of children who need someone to talk to.

In an elementary school, for example, when the library has open checkout time, the librarian may want to be present in that safe haven corner (see the discussion in Chapter 11 about the library as a safe haven) to lend children a calm and rational listening ear. In addition to finding opportunities to help individual children who need a listening ear, the librarian should also be aware that some children need the opportunity to both discuss and be listened to in their familiar social circle as a group or class (see the discussion later in this chapter about listening to and talking to children about disasters).

When major disasters occur, rarely is there just one issue that needs to be addressed. Rather, children and their families often find they must cope with multiple and often complex issues, many of which they have no previous experience with. On some occasions an event such as a natural disaster can create estranged or otherwise difficult living situations. Children may have to be separated from their family or a beloved pet (and sometimes that pet separation is permanent). When children are uprooted from their homes, their valuables and personal belongings may be left behind or destroyed.

The multiple issues surrounding displacement—loss of personal property, unfamiliar new living conditions, change to or limited transportation, unavailability of connecting with previous community members such as neighbors, or homelessness—often create enough complex sources of stress that a child may have difficulty verbalizing or identifying what is causing the greatest distress. Disasters can also cause unfamiliar financial constraints on a family that create multiple stressors. As the school information specialist, librarians may be the perfect resource for **connecting** children and their families **with assistive outside agencies, support networks, information,** or other services they need.

Following a disaster every individual involved needs assistance with hundreds of details. Many of these involve research. As the school's **research specialist**, the librarian may wish to remind the school and larger community of that professional role and promote the expertise a librarian has to help connect children and information, whatever that information might be. It may be as simple as a child seeking a replacement for lost valuables or helping a child find information to assist families in various recovery roles.

If large numbers of the school's impacted children have become displaced or homeless, or if they are residing in or seeking temporary shelter, the communication system previously in place may be disrupted. Allowing children to use library computers for personal email, the library telephone, or even the librarian's personal cell phone, as a **means of communication** for such things as checking in with parents when a student is in need of family assurance may be another service the librarian can provide. Granting use of communication tools not normally available to help a student feel safer will produce a grateful student response and potentially a long-term *positive* memory during an otherwise horrifically memorable time.

Research shows that next to immediate families or caregivers, the school plays the second most significant role in helping school-age children survive and recover from traumatic events. Just as children with a chaotic or abusive home life may discover school as the central place that helps them organize their day and provides stability, so too during disaster recovery the school might help children deal with trauma and move toward recovery. In the library, either as supplementary material, a single library lesson, or an all-school disaster recovery curriculum, teach new **organizational skills** children need to move on or to help organize their new lives.

Librarians certainly have firsthand knowledge about filters. Examples abound regarding both information access and filtering with which librarians must deal, and frustrations over a district's filtering policy that causes overfiltered Internet access when the librarian is trying to teach information skills is the universal example. Dealing with school boards, parents, and administrators in creating or changing policy around Internet access and filtering is another.

Librarians suffer encounters with upset parents who have found out their child has discovered inappropriate information while using the library computers. Even librarians themselves have to deal with filtered information—particularly in the primary grades—in a collection's books and magazines. But following a disaster, the librarian's stance on access and filtering may change, and the librarian may choose to be an **information filter** for unsolicited contacts, media, and public exposure. Helping young children obtain a developmentally appropriate yet undistorted understanding of events that have transpired may help these youngsters on the road to recovery.

One of the most important services the librarian can provide students when disaster or traumatic events have taken place is **literary diversion** from the event. The discussion in Chapter 10 about books that heal and books that help addresses books as bibliotherapeutic resources and explains the role literature can play in dealing with stress. The chapter also provides ideas for literary diversion. The appendix provides lesson plans librarians can use, some as diversionary tactics, others to help instill a sense of safety. As they work to create resources such as the temporary after-school activities like Recovery Reading Club or a Safe Haven Center to address whatever needs students have, librarians' diversionary offerings need be limited only by imagination, energy, time, and stretch.

Create Temporary Postdisaster Clubs and Activities

During times of recovery, members of the school community often draw on their own creative resources and inner resolve to create avenues of distraction, diversion, and renewal. The following list provides suggestions of temporary groups, meetings, or gatherings to offer to help children recover:

- *The Safe Haven Center*
- *Personal Mementos Memorial Service*
- *Poetry Writing Club*
- *Reading Recovery Club*

When outside professionals such as counselors and other mental health professionals are brought into the school community, as valuable as these resources are, some children find it difficult to share their concerns and personal stories with strangers, and the librarian can serve as a liaison. Children may feel comfortable accepting the librarian's advice when he or she delivers students into the hands of a trusting stranger, such as a social worker or mental health professional who has been brought in to help with coping skills. This may be particularly true in smaller communities where children have come to know their librarian as a community member and one of the significant trusted adults of the community.

Some of the acts for recovery do not require counseling services, recovery response personnel, or financial support. As one of the few educators who has access to and familiarity with all the children in the school, following a major disaster or crisis when every member of the school must work toward community recovery, the librarian may offer to help each class establish personal goals to guide recovery efforts. For example, as each class comes to the library for checkout, as a lesson introduction or prior to checkout, the teacher-librarian may offer a "check-in" time and, depending on results of that check-in, initiate an idea center or establish recovery plans in the form of goals. This can be initiated as an individual goal-setting activity or as a classwide, gradewide, or even schoolwide **recovery goal-setting**. This activity can help students focus on the future and recovery rather than the traumatic event.

Every team appreciates a cheerleader. A more abstract role the librarian may assume is that of the **school's optimist** or **cheerleader.** Engage students in productive, recuperative, optimistic activities. Display signs of encouragement and recovery (see SIDEBARS.) Distribute lists of uplifting book titles (see Chapter 10) and create displays of titles that encourage survival, recovery, and productive solutions. Find websites and other resources that provide encouragement and distribute these links to the entire school community, including families. Seek out stories of heroic deeds, selfless volunteerism, survival-themed stories, and literature of encouragement. Produce a temporary newsletter containing stories of local selfless acts, suggestions of encouragement, or necessary resources for faculty, students, and families.

On the first day back to school following a disaster or crisis, the librarian can be a **school "greeter"** standing in the halls to greet each student by name. If the librarian teaches in a school small enough or if he or she has been in the school long enough to know all students by name, this might be one healthy contribution to offer immediately following an incident.

In any school crisis some students are affected more than others. Also, some students will be identified as being at greater risk for negative outcomes related to the trauma. As a faculty member who has access to all students, the librarian may choose to volunteer for checking in with certain students during the recovery process. **Check in, double check, and check again**. Simply asking these students how they're doing or if they need additional support lets them know they are cared for and supported in their recovery efforts.

Baggerly and Exum, (2008) stress the importance of educators providing students with back to normal academics as well as adaptive coping strategies following a disaster, and he provides suggestions ranging from writing feelings down as therapy, to reading about happier times, to writing encouraging or "healing" letters to survivors.

The previous suggestions primarily focus on helping students recover. The librarian may also provide assistance to his or her peers. For example, when a suffering fellow teacher needs to be relieved temporarily, the librarian may volunteer to step in and take over a class.

SUMMARY

The various roles the librarian can play in disaster recovery need not be limited to the aforementioned. That is, as the information, literary, and research center of the school, the library offers a wealth of resources, and the school librarian can lead members of the school community toward healthy recovery efforts in whatever way the school community needs.

Motivational Displays

Small additions to the library facility may provide inspiration for recovery. Chances are that following a major event librarians will not want to spend precious fiscal resources on extras such as motivational posters. However, posters can be easily created for temporary displays of encouragement. Print the following quotations, each on an individual page, and display them next to the author's name with perhaps the quoted person's photograph.

"And will you succeed? Yes indeed, yes indeed! Ninety-eight and three quarters percent guaranteed!" **—Dr. Seuss**
"After a storm comes a calm." **—Matthew Henry**
"Be kind whenever possible. It is always possible." **—Dalai Lama**
"With the new day comes new strength and new thoughts." **—Eleanor Roosevelt**
"When we are no longer able to change the situation, we are challenged to change ourselves." **—Viktor Frankl**
"Optimism is the faith that leads to achievement. Nothing can be done without hope and confidence." **—Helen Keller**
"Things do not happen. Things are made to happen." **—John F. Kennedy**

"If you want to conquer fear, don't sit home and think about it. Go out and get busy." ***—Dale Carnegie***
"What you do today can improve all your tomorrows." ***—Ralph Marston***
"Our greatest weakness lies in giving up. The most certain way to succeed is always to try just one more time." ***—Thomas Edison***

Overcoming Obstacles at Any Age

Following a disaster librarians may also wish to display stories or create bookmarks about youth who accomplished amazing feats while overcoming adversity, for example, Anne Frank, who wrote The Diary of Anne Frank *at age 12 while in hiding from the Nazis, or Helen Keller, who was blind and deaf who with her teacher helped recreate sign language for more accessible communication for the hearing impaired.*

Displays for the High School Library

Though some librarians may feel that posting the following information is not appropriate for their school, others may find it helpful and appropriate to share with students facing adversity these anecdotes about famous people:

- *Comedian Jim Carey dropped out of school at age 15 to support his family and was homeless for a time before persevering to become a famous comedian.*
- *South African president Nelson Mandela began his life a cattle-boy tending herds and when at age 9 Nelson's father died, his mother took him to live and be educated far from the village he had known and loved.*
- *Johnny Depp lived in 20 different homes by the age of 15.*
- *Olaudah Equiano was kidnapped from his home in Nigeria and sold into slavery, where he suffered extremely cruel treatment. He later became an author, historian, and world explorer.*
- *At age 15 Charlize Theron witnessed her mother kill her alcoholic father in self-defense. She later left home and became a critically acclaimed actress.*
- *Jay-Z overcame an early life surrounded by drug culture, poverty, and violence to become one of the country's most popular rap stars.*
- *Mark Wahlberg, whose teen years were marked by violence, racism, and drug use, has worked to reconcile his earlier misconduct through his Mark Wahlberg Youth Foundation.*
- *Oprah Winfrey, who left home at the age of 14 after being molested by family members, gave birth shortly thereafter to her only child, a son who died in infancy. She has since become one of the world's most successful media personalities.*

Well Wishers, Worldwide

When vandals committed a horrific hate crime in the Talmud Torahs Elementary School library in Montreal, Canada, in April 2003, setting fire to the school library and destroying all but 25 of the 10,000-volume, multicomputer school library, to aid student recovery, officials created a notebook filled with emails from well-wishers all around the world. School officials wanted children to know that they were not alone, that others were thinking of them in their fire recovery. (Whelan, 2004)

Librarians: First, Help Yourself

On an airplane trip when flight instructions announce the airline emergency plan—"Should there be a drop in air pressure . . ."—at the beginning of every flight, they explain the importance and rationale for adults putting on their own masks prior to putting on children's masks. Adults must first be healthy, safe, and prepared before they can help children. So it is with disaster recovery. Use and share the following stress reduction techniques adults may employ:

- **Replan**: Following a disaster all adults in the school building should review their own emergency plan and change or create an up-to-date safety plan, if necessary. Interestingly, this can help provide adults with a sense of accomplishment and greater sense of security.
- **Write**: Just as children are encouraged to write down their thoughts and feelings in response to the disaster, adults too may find writing about their experience a healing process. In addition, writing down specific fears and possible solutions to resolving those fears can assuage issues and help in recovery.
- **List**: Simply creating a list of plans, strategies, or ideas on how to cope; ways to feel better; or personal favorites or rewards reassures an individual that he or she is working toward recovery. Additionally, disasters or crises often create multiple new tasks. Checking things off a list produces a sense of moving forward.
- **Breathe**: Learn deep breathing exercises, practice them, and, once mastered, use them as opening exercises with students prior to teaching a lesson. Breathing exercises are quick, are simple to use, require no equipment, and produce great results. Resources are readily found online for deep breathing, diaphragmatic breathing, and bellow breathing.
- **Be calm**: Learn other relaxation techniques.
- **Read**: Suggesting a librarian should read for pleasure is like singing to the choir, that is, seldom must librarians be advised to read. However, following a disaster so many tasks need to be accomplished that it is difficult for anyone to find time for those activities that bring reward. But do. Find time to read necessary information and resources for recovery, but also include pleasure reading as a necessary activity for recovery and diversion.
- **Reward:** Outside of school, schedule time for whatever it is that is most rewarding or relaxing and add it to the calendar or make reservations: bubble bath, massage, a comedy or classic movie, or a humorous or romance "escape" novel.
- **Create a safe haven**: Create that safe haven reading corner in the library if one has not already been established. If your personal life has also been interrupted by disaster, if possible, create a personal safe haven reading corner at home as well. And use it.

Helping Special Needs Students Cope

Addressing the individual recovery needs of special needs students may require a particularly patient adult. While classroom teachers of special needs students are skilled in helping these children deal with daily life struggles, during a major incident special education teachers may be overwhelmed with the tasks before them. Further, while their training may provide tools for day-to-day special needs' struggles, these teachers may need additional help and training related to disaster recovery. Each special needs child's response to disaster and the corresponding recovery and time frame will be unique, may be less predictable, and may be further complicated by individual needs.

> There may be ways the library facility and librarian can help that seem minor but to individual teachers and students may provide a great mechanism for coping. For example, individuals in the special education classroom may feed on each other's fears and incite each other to upsetting responses or behaviors. Therefore, some individual students may just need to get away or be separated from their classmates. Librarians may help by providing a quiet, calm area for a student to come and relax, perhaps for silent reading.

At such times during disaster or crisis recovery, even if special education teachers have not called on the librarian before, such assistance may be greatly appreciated. For example, the librarian may be available simply to read to students. Or the librarian may provide the special education teacher with literature to read: calming or gentle stories, poetry, or samples of biographies of people who have endured great obstacles.

Providing reading material may be a service the librarian can provide for all faculty, but it may be particularly appreciated as an unsolicited service for special needs students. Seek out and offer to provide these therapeutic reading resources (see Chapter 10) not just to special education teachers but to all classroom teachers in the building. Finally, as with any classroom recovering from disaster, there may not be enough adults to go around. The librarian may be able to provide the special education teacher with another helping hand, a listening ear, or a means of distraction from the event.

LISTENING TO AND TALKING TO CHILDREN ABOUT DISASTERS AND SCHOOL CRISES

Our schools should and must be one of the safest places in society. Now it is time for another, quieter heroism in our schools—the courage to move forward, to continue to teach and lead children, and to take smart precautions to minimize the risk of future tragedies. The challenge after Sandy Hook of trying to explain the inexplicable to young children is very real. But teachers, principals, and school staff are also feeling fragile. They want to be sure kids get the mental-health services they need to prevent future tragedies.

—U.S. Secretary of Education, Arne Duncan, December 19, 2012

As the American Academy of Pediatrics advises in the opening statement on their advisory page, *Talking to Children about Disasters*, "Children can cope more effectively when they feel they understand what is happening and what they can do to help protect themselves, families, and friends" (American Pediatrics Association, 2014). Following a major disaster or crisis, most children need multiple opportunities to talk about what has transpired and, perhaps more important, discuss and understand the plan for recovery.

When school-age children have experienced devastation from disaster, whether human-caused or natural disaster, they experience a heightened sense of vulnerability that can be moved toward recovery through honest dialogue and sharing. While professional organizations and publications do not agree on whether it is necessary for adults to initiate those questions and needed discussion or whether this should be student led or initiated, one belief that all share is that adults need to look for the best moments for discussions and provide opportunities for dialogue in order for children to verbally vent after they have been exposed to a traumatic event. Start discussions by asking children what they know already and what they've heard or seen.

Adults in the school community are encouraged to provide ample appropriate opportunities for children to express their feelings and thoughts about events that have transpired. It is important for children to understand the event itself and to be able to discuss their perception of the event as well as their perception of what will happen in the future, both related to the event itself and to recovery efforts. In the library there may be unique opportunities to allow classes to discuss events and subsequent recovery.

Model Productive, Positive Discussion

Following such emotion-charged times it is critical to establish an open, democratic discussion with clear standards for open dialogue. During discussions and when providing answers, particularly to difficult questions, librarians should take the opportunity to model productive discussion techniques. For example, they may wish to share their own concerns and how they are dealing with them. Or librarians may share the fears or concerns they had immediately following and during the crisis and how they dealt with them (being cautious, of course, not to burden children with their own or additional fears). These examples reassure children that their fears are normal and universally felt.

As difficult as it may be, to seek resolution and recovery for all students, an atmosphere of open and honest dialogue must be established. This means that, within reason, discussions must allow for disagreement and the airing of differences in perspective and points of view. Children who are told by trusting adults that their ideas and beliefs are wrong, particularly if this is in opposition to the child's familial, religious, or cultural beliefs, are likely to disengage.

One effective strategy to promote positive ways of coping is to end or summarize discussion periods by reassuring children. Provide information about how they are protected following a disaster and the safety measures that have been put into place following the event.

The Media

Media coverage can provide opportunities for critical discussion. While it is advised that younger children in particular have limited access to news media coverage, it's impossible to shield all children from the omnipresence of the media when a major incident transpires. It is the adults' responsibility to provide clarity with regard to what is being presented and to put the news coverage into perspective. Multiple opportunities arise for librarians to engage in dialogue related to media coverage and disaster recovery.

> Librarians may take the opportunity with older students to use media coverage in a lesson or discussion of fact versus opinion, media bias, positive versus negative reporting techniques, or the dramatization of news coverage for effect, hype, or increasing numbers of viewers. With younger children, the librarian may simply help clarify misconceptions about media coverage such as explaining that repeated stories are not repeated events; media reporting presents the most dramatic events; and time frames and geographic locale may need to be clarified.

> Librarians may also wish to provide alternative stories to contrast media hype or dramatization of events. For example, finding and presenting a story of a rescue, a selfless volunteer, or other demonstration of community strengths and steps toward recovery may provide a welcome relief from more dramatic media stories. This can be particularly effective if the alternative stories involve the disaster at hand.

PERSONAL STORY: Sharon's Story

When I first began teaching about 35 years ago in a small rural K-12 school, no one had emergency preparedness plans or communication plans detailing how to handle the media during a critical event within the school. One particular eighth grade girl was very advanced physically and had beautiful long blonde hair that looked like Farah Fawcett's, who was a television idol of that time. This girl, "Patty," was invited to the prom by a high school junior (there was no school rule preventing younger students from dating upperclassmen). Patty's parents gave her permission to go to the prom partly because she had an older sister and would double date with her. This was a small community where all the kids had grown up together and had known each other forever. They were in community clubs and organizations together, and the dating ritual was somewhat informal, like a group date.

However, another group of eighth grade girls was very jealous of her, particularly her hair, and would taunt her with "mean girl" junior high behavior. One day after school a colleague and I were monitoring the kids in the hall as they were at their lockers gathering what they needed prior to leaving for the day, when I suddenly heard a horrible scream. Fortunately it was a very small school and I saw immediately that Patty was bent down on the floor in front of her locker, screaming. I could see that her hair was smoking! The basketball coach, who happened to be the one on hall duty with me, picked her up and we both ran to the gym, down a flight of stairs, and literally threw her in the shower. I stepped in with her and began to wash her hair, which came off in big clumps in my hands.

Apparently one of the girls had stolen some acid from the science lab and had thrown it on Patty's hair as she bent down to pick her books up out of the locker. The deadly acid could have eaten into her spinal column or brain stem from the place it was thrown. The only thing that saved this girl's life was our quick thinking to throw her in the shower and the fact that she had on a heavy coat up around her neck.

The girls didn't really understand the severity of what they had done. They were not trying to kill her—they just wanted to get rid of that beautiful hair!

The acid throwing crisis was bad enough, but things escalated rapidly and became much worse primarily because was there was no plan for communicating to the community and soon enough to the media. The principal was a publicity hound. He immediately began issuing statements, telling everyone who would listen, and the story spread like wildfire. It soon became a story of national interest when it was highlighted on Good Morning, America *within a day.*

The media attention did great damage to the girls and the school, which got a terrible reputation that took years to live down. The girl who had been hurt had years of surgeries and skin grafts, which was bad enough without the media hounding her. The media hype escalated the story such that the perpetrators were accused of things they really did not mean to do. The girl who threw the acid had to change her name and move to another state. All because, in a school where the principal liked to see himself on TV, there was no plan for effective media communication response to a crisis.

—Editor Sharon Coatney

Student Questions and Concerns

Children and their parents depend on schools to help children understand and deal with important events in their lives and communities. If we avoid dealing with important issues in schools or imply to children that we do not think they are capable of discussing difficult topics or that it is inappropriate to talk about these events and their associated feelings, we risk children seeing schools as irrelevant to their personal lives and unable to deal with difficult issues. We risk teaching them that these events and the many lives impacted by these events are unimportant or inappropriate to discuss. We risk implying that there is something wrong with them that they are unable to deal with such events and the associated feelings.

— Salamone, 2006

Invite questions when discussing both the event and recovery. Misinformation or misinterpretation, such as that produced through media hype, often causes erroneous assumptions. These assumptions can provoke anxiety, particularly for the youngest viewers. When they are able to ask questions, children are granted an opportunity to express their fears as well as an opportunity to discover they are not alone in their feelings. Often when children begin to ask questions of trusting adults, such as their librarian, they discover that some of their fears are unfounded and they may also be the same fears that their classmates have. Understanding that their fears are universal helps on the road to recovery.

Librarians must use their professional judgment, however, to determine when the questioning and discussion reaches a point of being unproductive or even harmful. On occasion the questioning devolves into dwelling on difficult events or feeding rumors. On rare occasions a student may even create a completely false story, rumor, or report to get attention or sustain the drama. When this happens librarians need to reorient children to other ways of responding or other pursuits. For example, if some students still have legitimate questions and need answers but the questioning has taken a disturbing turn, the librarian may suggest having students write down their questions for later. This also might produce questions that some children still have but are uncomfortable sharing with the whole group.

Librarians can then decide the best way to respond to the written questions. If it is obvious these questions are serious enough that they need to be addressed in a formal way, the librarian may need to begin the referral process. This may also be an opportunity for small group discussion.

Tough Questions

Tough questions will come up. According to the About Our Kids organization, "Contrary to parents' fears, talking about violent acts or threatening events will not increase the child's fears. It is very important to engage in an open discussion about children's feelings, fears and worries. Avoiding discussion of scary feelings may be more damaging than talking about them. Even children as young as four or five know about violent acts, but not all children may know how to talk about their feelings and concerns" (AboutOurKids.org, 2011). Be mindful that under extreme circumstances such as a major disaster or all-school crisis, children often behave uncharacteristically. Therefore, as tough questions arise educators must be careful to respect and expect an odd assortment of responses in the form of questions.

Do not assume what children fear most; children's questions and personal fears may be surprising. Further, when it comes to recovery measures, no personal fear or honest question should be criticized, minimized, or ignored. Children need the freedom and opportunity to explore the issues, ask the tough questions, and find ways to cope. Helping children understand and master their fears will expedite the road to recovery. The best way for this to happen is by providing answers when possible but at minimum providing the opportunity to ask and discuss tough questions.

Dealing with tough questions that relate to personal beliefs can further complicate discussion, even questions that initially had personal recovery as a goal. Within any class or group of youngsters, children will have opposing beliefs—often based on family culture—especially related to school violence issues. For example, there will be children at both extremes of the complex issues regarding First Amendment rights and guns, particularly assault weapons. Students who are (or whose families are) strong gun advocates may feel guilt, hurt, or anger when it comes to school shootings. Further, if these students are vocal in sharing their beliefs, they may become victims themselves of displaced anger from fellow students or even adults in the school.

Following a school shooting educators too may find it difficult to remove personal and emotional bias when it comes to guns. Recognizing and acknowledging that emotional responses are normal and heightened after a tragedy may help assuage negative emotions and move toward recovery. During discussions when the response to an act of violence turns negative or hateful, it is the responsibility of the educators to help children acknowledge their own pain and anger as well as their personal biases, and encourage them to find a positive means through which to channel a response. Likewise, educators must examine their own biases and be mindful of them when addressing tough questions, modeling democratic ideas, and hosting response team members. As much as educators may wish to help students move toward recovery, when discussions warrant professional expertise, call on the school's mental health professionals to take over.

Chapter 10 provides primarily fiction literature to share with students following a disaster. Librarians may wish to read some of these stories as a springboard for allowing students to discuss their response to disaster or crisis. These may also provide the impetus for students' own literary inquiry. That is, while they might not have thought of such, with the introduction of a story of rescue or recovery, children may request additional suggestions for reading similar stories. Stories of heroic acts and stories of recovery can provide a sense of hope as they provide examples of recovery and safety.

PERSONAL STORY: Sandy Hook

Following the tragedy of Sandy Hook Elementary, I found myself aching and angry at every acquaintance I knew who supported gun rights. For weeks afterward I read everything I could get my hands on to arm myself with facts and statistics on the violence brought about by guns in our country. I ranted in emails, posted on Facebook, wrote letters and cards to legislators. I rallied against violence: television violence, movie violence, video game violence. I wanted to lock my first grade seven-year-old grandson in a bell jar for protection. For a time I wanted be around only like-minded individuals who knew "what was right" and who would agree with everything I believed about guns and violence. I did not want time to heal, rather I wanted the world to change; I wanted our country to stop the violence and turn away from the culture of Gun Love.

After a year, on the anniversary of the tragedy of Sandy Hook Elementary, and well into writing this book, I began to reflect on my own frame of mind. I examined my response, came to admit my personal biases as just that (rather than "what was right"), and to some extent, I have recouped a prior more objective stance. But not altogether—I am not where I was prior to the tragedy of Sandy Hook Elementary. I know this is a normal response.

My granddaughter's teacher, Kate Sipe, said it well, "Sandy Hook changed everything." Today, and through the therapeutic act of researching and writing this book, I am able to acknowledge my stance, my beliefs; and I use that initial research on violence and guns in this country to advocate for the overall safety of our schoolchildren, for sensible legislation that preserves most individuals' constitutional rights. Perhaps this is what we can hope for most when helping our children recover from natural disasters as well as acts of school violence: a move to reasonable action.

—Christie

PERSONAL STORY: Oklahoma City Bombing

In April of 1995 I had the honor of responding to the Oklahoma City bombing as a task force leader of Washington Task Force 1, one of the 28 national urban search and rescue teams developed by FEMA to support local and state responses to collapsed structures. We arrived on Day 5 after the bombing to relieve one of the FEMA teams that had been there since Day 1.

For the next six days we worked 12-hour shifts digging through the rubble of the collapsed Murrah Federal Building. The bomb had killed 168 persons, including children who had been in the building's day care center, federal employees, and citizens who were in the building on business. Hundreds had been injured.

We worked with the Oklahoma City Fire Department, the FBI, the National Guard and many other responders in recovering the dead. It was backbreaking and heartbreaking work.

When our relief arrived we had a day to rest before returning home. We were invited to come to a local elementary school to talk to the children. Everyone knew someone who was affected by the explosion: someone who had lost a parent, a loved one, or a friend. Oklahoma City was a close-knit community, and the students were especially affected by the disaster.

We spoke to the children about the work we did, how our search dogs were trained, about the equipment we used in rescue, and how honored we were to help families and the community reach a measure of closure. It was a very moving experience for us. I remember when we were done, a child came up to one of our rescue specialists, who was a sheriff's deputy in Pierce County, Washington, and said, "When I grow up, I want to be just like you."

Our hosts, the teachers at the school, sent a letter to me expressing how much the students and staff appreciated our coming and how it had helped in their recovery, by just talking about it with us.

—Bill Lokey

Dealing with Personal or Conflicting Beliefs

For some educators, particularly with regard to acts of terrorism such as 9/11 or school violence such as the tragedy of Sandy Hook Elementary, removing personal biases or conflicting beliefs may be difficult.

While it is the goal of every educator to move students toward recovery following a disaster, some adults struggle themselves to make sense of horrific events, especially when their personal beliefs lead them to feel the event could have been prevented. Librarians and teachers holding strong beliefs that conflict with those of other adults or students at the school are cautioned to remove themselves from unproductive dialogue or worse, blame. This is particularly true if victims or perpetrators were a part of the school community. If the events are too personal or dialogue and discussion are too difficult, let mental health professionals assume this professional role.

The Urban Search and Rescue Team from Washington State works in the rubble of the Murrah Federal Building after the bombing in Oklahoma City in 1995.

Dealing with a School Member's Death or Suicide

Many of the suggestions, resources, and ideas in this chapter address issues that schools encounter when they are struck by a major natural disaster or all-school crisis. Many of the ideas suggested here can also be applied in the school that has been impacted by the death of a single student or staff member that is not the result of a major incident.

The American Academy of Pediatrics (AAP) has an excellent website that provides step-by-step suggestions for helping students deal with the death of a classmate or staff member. In these instances, the school crisis team in concert with school counselors and local mental health professionals are called upon, and often a school district with multiple school buildings will send all counselors to a site that has been affected.

Librarians should be aware of a few unexpected potential posttrauma responses. Occasionally, some time later, students will come upon personal effects or other reminders of the student or staff member they lost, such as the name of the deceased on school property or a textbook. When this happens, unexpected reactions may occur. For example, some students who had ambivalent or even negative feelings toward the deceased may respond in ways that seem insensitive. This may incite other students who are still grieving to display a more highly emotional response. When such incidences occur, members of the counseling or crisis team should immediately be called in again. One of the goals of all staff members is protecting students from retraumatization.

Acknowledging Anniversaries

Whether it acknowledging the death of a single student, teacher, coach, or family member or acknowledging a larger event such as a hurricane or 9/11, event anniversaries are to be remembered and memorialized. These anniversary events are rites of passage and help in recovery. They are not to be ignored, diminished, or excluded when children or other adults in the school wish to acknowledge anniversary dates.

> Children and adults may have very different needs around the time of the anniversary and hold very different opinions about how best to acknowledge the anniversary—Indeed, there is no one best way to mark the anniversary. Children who've experienced a personal loss may prefer to spend today alone or in the company of their family or friends or may instead want to honor their loved ones through a public event or activity. (Salamone, 2006)

The librarian may wish to volunteer use of the library facility as a location for memorializing the event. If the librarian is serving as one of the hosts, consider that one of the most valuable contributions to make lies in focusing on positive changes that have transpired since the event occurred. This will help all involved look toward a positive future as greater recovery is made. Additionally, focusing on the fact that while the event itself may not have been preventable or planned, all aspects of memorializing demonstrate the importance and plausibility of planning and moving forward.

In addition to the anniversary of a fateful event, holidays and other special occasions, particularly those occasions on which individuals were directly affected, might be difficult times; for example, the first birthday of a lost member of the school community. In anticipation of these occasions, educators must understand that there exist no "absolutes" with regard to appropriate recognition. With those children directly impacted, librarians and others should keep an eye out for recurring or regressive response behaviors or other symptoms of retraumatization. Educators may help younger children understand why they are feeling as they do. Again, simply lending an ear, an open door, or resources may be the most respectful response.

An event acknowledging difficult times, while important, is rife with potential vulnerabilities. Therefore, planners and host must recognize the importance of ensuring that the event is well structured. Further, while events may be organized and well structured, plan for flexibility and the unexpected. Know that well-intentioned adults may divert activities from planned events.

When the event takes its place in history and an ordinary day is just that, whether individuals recognize it or not, the school community has turned the corner toward recovery.

SUMMARY

When disaster strikes, as a member of the school community, the librarian is jettisoned onto an emotional journey of reconciliation and restoration along with all other victims and becomes a passenger on the journey of recovery. No matter how traumatic the event, time has a way of helping with recovery. A day comes when an ordinary event is seen not as symbolic, a diversion or a relief, but just that: an ordinary day. Few individuals who have experienced a large-scale disaster or horrific event deny the value, importance, and beauty in experiencing the calm of an ordinary day.

Chapter 10

Books That Heal, Books That Help: Using Literature in Disaster Preparedness and Recovery

As we face uncertain futures, children must know that they can find not only comfort, but hope in books.

—Bedford, 2007

Librarians can offer a multitude of services to children that incorporate the healing power of books as well as the magic of escaping into a great story. Whether providing a one-to-one suggested book for a child in need, all-class book promotions, or a library read-aloud, every opportunity to bring stories into the lives of children can be an opportunity to keep them safe or help them heal.

BIBLIOTHERAPY

For years school librarians have used bibliotherapy strategies to serve patrons in various ways. Constructive contributions of the roles bibliotherapy plays in children's lives include the power of literature to deepen understanding; to provide children with the knowledge that they are not alone in their experience and feelings; and to help them heal and cope with obstacles, adversity, and great calamity.

> "Validation of a child's experience, thoughts, feelings, and actions. The truth of events for each of us is found not just in what happened, but also in how we felt and thought about it then, how we feel and think about it now, and the reactions of significant others" (Allen et al., 2012).

Allen and colleagues propose that librarians become involved in what they refer to as "crisis bibliotherapy" to offer children what librarians do so well: an opportunity to share a story or two in the aftermath of a disaster and discuss the event within the context of the story. In addition, during a critical event, for example, when children are in lockdown, being able to make references to stories they have read or are familiar with may provide them with both calming and problem-solving opportunities. Stories provide opportunities to bring shared experiences together and offer a stage for making meaning from them. Using stories in this therapeutic manner, especially with the youngest readers, can bring a critical event into a new light and serve as an opportunity to use books to heal, at the point when children love stories most of all, that beginning-to-read stage in their early reading life.

Stories can also prepare students for *potential* disasters. "The more familiar children are with natural and human disasters, the more proactive their view of the world" (Zeece, 1998). Sharing well-constructed stories with disasters as background can provide valuable information and serve as a predrill activity for disaster preparedness.

The term "bibliotherapy" has evolved from a clinical term first defined in more general terms to one that now describes a variety of interactions between readers and books that help readers deal with critical issues in their lives. Some professionals in the field, however, believe that implicit in the application of bibliotherapy is the formal acquisition of psychotherapy skills.

It is not suggested here that school librarians act as counselors or therapists for children in preparation for or in recovering from a natural disaster, but the capacity for literature and stories to help children feel safe, calm, and not alone lies within the realm of the librarian's role. Therefore, to focus on the power of the story itself, while there may be bibliotherapeutic references herein, more general terms such as "book therapy," "books that help," or "books that heal" will be used to describe ways in which librarians can help children find solace through books and reading.

Librarian: The Overlooked Link

One of the most important functions librarians carry into the role of literacy support is to match the literature being read with what is important in a child's life, and there is likely no time more relevant in addressing the emotional events that happen than when children are in the throes of disaster or crisis recovery. This can be particularly cathartic because at the same time that reading stories with relevancy to what's around them provides abreaction, it has too a distancing effect that allows children to replace their too-personal concerns and fears with those of the character of the story and may afford alternative ways of responding to an event.

Books may help children suffering from survivor guilt as well, that is, suggesting healing books to a child who has experienced an event in which one or more of her classmates has lost his or her home, suffered great anguish, or not survived. This, at any age, can be one of the most traumatic and enduring experiences from a disaster. Providing a story that has a character with whom a child can identify may bring some solace and comfort and provide an avenue for discussing feelings.

MAKING SENSE OUT OF CRISES: BOOKS AS THERAPY

How might the school librarian apply such book therapy or use books to help children following a disaster? Schoolchildren, particularly younger children, who have experienced a traumatic event often have difficulty verbalizing their response. Younger children also may not be developmentally equipped to understand the complex nature, results of, or personal implications of these traumatic experiences.

Providing children with stories, particularly younger students whose maturity may dictate unique needs and limited understanding, with characters who must face similar adversity and story lines to which they can relate, offers them a chance to view challenges from someone outside their own crisis. This is the power of literature to provide a sense of calm and safety.

Library's Role in Dealing with Disaster

In "The Power of Story," Allen and colleagues (2012) provide multiple suggestions for the bibliotherapeutic role libraries can play—both during a disaster and after—in recovery:

- *A place of safety*
- *Productive action*
- *An atmosphere of calmness*

- *Validating experiences, thoughts, feelings, and actions*
- *Vicarious problem solving*

PERSONAL STORY: Horton Hatches the Egg

One of the first experiences that demonstrated to me the power of story was during my first year as an elementary librarian in a high-poverty, high-crime, urban school where students who lacked the requisite 10,000 hours of emergent literacy were incredibly street smart and ready-tough before even entering kindergarten.

One morning as I was reading Horton Hatches the Egg *to a class of first grade students, I came to that tense scene in which Horton is confronted by the malingering absent mother, Mayzie.*

As I read the lines spoken by Mayzie—"But it's mine!" screamed the bird, when she heard the egg crack. (The work was all done. Now she wanted it back.) "It's MY egg!" she sputtered. "You stole it from me! Get off of my nest and get out of my tree!"—one intense young boy jumped up and shouted, "No! She can't have him! She left him! No!"

I tried to soothe him, calmly saying, "Let's finish this story; I promise it will be okay."

Suddenly he realized that his outburst had revealed a bit too much about his inner anguish. He sat down, embarrassed, and mumbled something disparaging about the book.

Later, curious about the drama behind his extreme response, I went to the school counselor and asked about his personal story. Not surprisingly, the school counselor reported that the young boy was being raised by his grandmother after having been abandoned by his own mother, a woman with severe drug addiction; no one knew her whereabouts.

Prior to this, I had not heard of the term "bibliotherapy," but this experience prompted me to do more reading about the power of story in children's lives. And while there may not be a lot of empirical research on the impact stories have to help children heal, stories abound and librarians often witness examples such as this, giving credence to the importance of the librarian's role of matching a child with stories of comfort, solace, and identity.

—Christie Kaaland

READING ALOUD TO CHILDREN

Each reader brings individual experience and understanding to the story and takes away what seems significant.

—Hunsburger, 1985

When children experience a crisis, sometimes it is helpful for the librarian to match individual students with particular stories. Other times children need to hear stories out loud and find solace with their peers who have also experienced a shared challenging experience.

From the age-old art of reading a child to sleep to such contemporary literary-political campaigns by legendary reading advocates such as Stephen Krashen and Jim Trelease who tirelessly champion reading aloud, librarians just *know* firsthand the benefits of reading aloud to children. This is particularly important with children trying to make sense of a crisis or challenge brought on by disaster. Let them know that the library is that safe place they can go and be swept away by a great story.

When reading gentle or soothing stories for the purpose of escape or solace, librarians who consciously use a calm, soothing, even meditative voice can greatly comfort a child's consternation.

USING BOOKS TO HELP OLDER STUDENTS COPE

Well-written fiction exposes readers to challenging scenarios that can serve as grist for processing life's dilemmas and for considering alternative solutions.

—Larson & Hoover, 2012

It is likely for books to be used in elementary schools to help readers heal, and older students can also find solace in the pages of a book. For example, Karen Gavigan (2012) suggests connecting teens with appropriate graphic novels or comics to help them address challenges in their school lives. Pollack (2006) proposes looking for titles that provide examples of coping skills, empathy, and healthy and creative resolution for older readers who are considering issues related to bullying, a common thread in school threats of violence. Prater, Johnston, and Dyches (2006) say that "younger students might be more open to suggestion[s] of a specific book . . . [and] older students generally like to be given a choice," so displays, book trailers, and general book talks can be a good way to offer books to older students as they work to heal following school crises.

INCLUDE HEALING BOOKS IN THE SELECTION POLICY

One of the first documents that I read was by Vandelia VanMeter [1991.] She found that of several critically-sensitive needs areas that kids should have information about, students could not access the resources efficiently or readily because they were either a part of the professional collection and/or not accessible through regular bibliographical control systems. If her findings are true today, one of the first things we need to do is to put our own house in order with a good selection policy statement, collection and curriculum evaluations, and input from teachers.

—Barron, 1997

When in library school, future librarians are often required to create a selection policy for the imaginary school of their future career. This policy serves to consistently remind librarians about what they feel are critical considerations with regard to adding to a school library collection that aims to serve students and staff.

Consider adding themes such as violence prevention and school safety to the library's selection policy. Look for the following:

- Books with characters who face disasters and resolve their problems
- Titles about various natural disasters, both fiction and nonfiction
- Characters who role-model uplifting bravery
- Fiction topics for the natural disaster specific to the needs of the region
- Books with themes such as recovery and overcoming challenges
- Comfort stories

The next section of this chapter includes annotated collections of multiple topics related to disaster preparedness and recovery.

Selection Criteria for Books Related to Disaster Recovery

Consider the following criteria when selecting books to recommend to children following a school crisis or natural disaster event:

- *Plot demonstrates respect for the readers' tragedy*
- *Characters deal authentically and honestly in their disaster response*
- *Plot provides realistic resolution to a disaster or crisis*
- *Author shows respect for diverse cultural responses to disasters or crisis*

- *Illustrations support the tone of the disaster or event*
- *Plot offers readers constructive ideas for coping with crises*

For schools that have been subjected to a major crisis, librarians may want to limit purchases of titles, even popular books, that show unhealthy responses to crisis.

Kids' Four Greatest Fears in Disasters

When children are faced with a disaster, certain universal fears drive their response and actions. Consider finding books that help students address these fears:

- *It will happen again*
- *They will experience injury or death*
- *They may be separated their family*
- *They may be left alone*

SUMMARY

While stories and literature have the power to carry readers away, stories can also cause readers to revisit painful memories. Whether children request titles with plots of shared experiences or librarians take it upon themselves to match child to story, it is always appropriate when recommending a book to say to the reader, "If you don't like how this story makes you feel, close the book and bring it back to the library, and we will find a better book for you."

The rest of this chapter provides an annotated guide to picture books that address various subjects relevant for younger readers. All contain disaster preparedness and crisis recovery topics and themes. Specific themes include those associated with natural weather disasters, children as role models of bravery and overcoming fear, stories that comfort, and dealing with associated challenges such as separation and relocation. Included also are stories to help build knowledge of safety and preparedness.

The annotated guide contains primarily fiction picture books for several reasons. Picture books can be used as read-alouds in one sitting/lesson. Nonfiction books on subjects such as natural disasters or historic events (e.g., 9/11) are easily found with a quick subject search. It is more likely that librarians will have greater opportunity to select picture books to read aloud in a variety of instances to younger readers, for example, as an activity during drills or shelter-in-place situations, as library lessons, following weather-related events, and during the healing process following a major disaster.

> We cannot know what lies ahead for our children, just as we could not know prior to August 29, 2005, how our world would be irrevocably changed. We do know that children will always need good books and that reading will always have a place in their lives. (Bedford, 2007)

CHILDREN'S PICTURE BOOKS: AN ANNOTATED BIBLIOGRAPHY

> *I know that books alone will not heal all the wounds inflicted by this storm. However, I believe in the power of books to enrich lives under the most dreadful circumstances.*
>
> —Bedford, 2007

Selection Criteria and Rationale

Current trends point to greater integration of nonfiction across the literacy curriculum, particularly with many states' implementation of the Common Core standards. Nonfiction books and online sources are critical to informing students about safety and weather, and many excellent nonfiction titles are published for the purpose of providing that information. It is not the intent here to dismiss the importance nonfiction plays in the collection of titles for safety and disaster preparedness, but nonfiction titles are readily accessed through a catalog subject search.

In addition to informing students, books can also address the needs that students face in a crisis or the complex issues that come with experiencing a major natural disaster. Sometimes a child just wants to read a story whose dynamic character has experienced something similar. These titles are more difficult to find and cannot always be accessed through a simple search. For that reason the titles in this chapter are mainly fiction and were gathered based on the criteria listed in the following sections. Some titles were selected based on subject (e.g., fire, flood), whereas some more difficult to find themed titles include children as role models bravely facing disasters and the importance of community in recovery.

For brevity, the titles are nearly all picture books and most, but not all, are recommended for elementary students. Most often educators are able to share picture books in one reading or one lesson. Picture books can provide that opportunity within the library setting.

Arrangement of Titles

This annotated bibliography of children's picture books is divided into two general sections. The first section has fiction picture books related to natural, weather-related disasters. These are titles librarians may wish to read or recommend to teachers or individual students to learn about as well as to identify with characters who have experienced similar conditions. Librarians and teachers may wish to read these titles *prior* to students experiencing a disaster. The general topics are arranged alphabetically by disaster topic and within each category by author.

The second section provides books that may be used to comfort those dealing with fallout from disasters, natural or human caused. They are themed around issues that students experience following a crisis or disaster such as separation, fear, and isolation, as well as themes to uplift and comfort.

Because of the gravity of these topics and themes, deliberation was given to include primarily books of strong literary quality. Consideration was also placed on publication date because some books may be considered for purchase, although a few older classics are included. Titles that are starred are high-quality books this author highly recommends for disaster recovery or emergency preparation, whether for healing qualities, relatable stories, or informative content. Finally, great attention was also placed on representing as much cultural, racial, and ethnic diversity as possible.

Drought and Heat Wave

Most of the literature that carries stories of difficulties with water shortage, drought, and heat wave, understandably, is from outside the United States. Teachers and librarians may lead a global-focused discussion about world problems related to drought and water shortages.

*Aardema, Verna. *Bringing the Rain to Kapiti Plain.* New York: Penguin, 1992. (Gr. K-3)

While watching over his herd on the African Savannah, Ki-Pat repeats a refrain that he hopes will bring rain to the drought-stricken plains. In colorful illustrations of those golden plains and repetitive rhyme, this cumulative tale brings the African veldt, both plants and animals, to life.

Chang, Grace. *Jin Jin and the Rain Wizard.* Brooklyn, New York: Enchanted Lion, 2009. (Gr. K-2)

Jin Jin, a mischievous little Chinese dragon, has had too much fun at the rice festival and decides to stop his water-breathing ways, which in turn causes a severe drought. He is sent on a journey through time to visit his drought-stricken ancestors, where he learns an important lesson about wastefulness and conservation.

Conway, David. *Lila and the Secret of Rain.* London: Frances Lincoln, 2011. (Gr. 1-3)

To relieve her Kenyan village of its terrible drought, Lila decides to visit the sky to tell the saddest story, hoping the sky will weep, as she's learned it might from her grandfather's stories. When she tells of the village drought, the desired results occur, and she returns home to find her village celebrating a downpour.

Cowley, Joy. *Singing Down the Rain*. New York: Harper Collins, 1997. (Gr. 1-3)
All the townsfolk are getting on each other's nerves as an incessant drought zaps everyone's positive energy. Then an energetic woman roars into town claiming to "specialize in rainsongs" and begins a lively chant, encouraging everyone to join her. It works! Raindrops fall and spirits rise.

DePaola, Tomie. *The Legend of the Bluebonnet*. New York: Puffin, 1996. (Gr. 1-3)
This Comanche legend presents a tale of unbridled selflessness when a young orphan girl sacrifices her only possession, a warrior doll, to the Great Spirit in order to bring an end to the drought that is blistering the land. Her good deed is rewarded when, where ashes from the doll have been scattered, a field of bright blue flowers bloom.

Faundez, Anne. *The Day the Rain Fell*. London: Transworld Publishing, 2010. (Gr. K-2)
This African creation story presents Lindiwe and her daughter Thandi, who come to earth to check on the land. They discover severe drought and the animals of the African savannah languishing. So Lindiwe brings rain. As Lindiwe creates pots wrought from the red clay and water, Thandi designs beads to which all the animals contribute their bright colors, from flamingo pink to ant red.

Hamilton, Virginia. *Drylongso*. New York: Houghton Mifflin, 1997. (Gr. 1-5)
Following a long drought, Lindy and her family rescue a mysterious boy during a dust storm. He brings mystic and prophetic words along with the promise of water. Jerry Pinkney's lavish and unique illustrations bring the dusty desert to life.

*Hesse, Karen. *Come on, Rain!* New York: Scholastic, 1999. (Gr. K-3)
Tessie and her mother struggle to endure the heat; then Tess, from the fire escape, sees the clouds roll in for blessed relief. First, just the girls—Tessie and her friends—rain dance in the street, and soon all the mothers join the dance. Readers will feel the "rain cleansing" in this female-filled, dancing-in-the-street urban delight.

Kessler, Cristina. *My Great Grandmother's Gourd*. New York: Scholastic, 2000. (Gr. 1+)
When a Sudanese village installs a new water pump, villagers hope to put old water-gathering customs aside. However, Fatima's wise grandmother recognizes the value in keeping the old along with the new and when the pump fails the water grandmother has gathered is a welcome relief.

Lind, Michael. *Bluebonnet Girl*. New York: Henry Holt, 2003. (Gr. K-3)
This traditional Comanche legend tells how the people are punished for their greed, causing drought to fall upon the land. The solution requires everyone to sacrifice into the campfire that which is most valuable to them, and one young girl places her doll in the flames. The next morning rain comes, and the doll's feathers bring a field of bluebonnet flowers. (Pair with DePaolo's book of the same Comanche legend, *The Legend of the Bluebonnet*.)

Stojic, Manya. *Rain*. New York: Random House, 2009. (Gr. P-3)
Readers can just feel the parched ground of the drought-weary countryside as the animals of the African veldt eagerly await an impending storm, each chanting, "The rain is coming." Each uses its keenest sense to predict the rain. Illustrations portray each weather condition as the gorgeous reds sizzle in the heat and then grow gray and more daunting with the storm clouds.

Walter, Eric. *Hope Springs*. Toronto, ON: Tundra Books, 2014. (Gr. 1-4)
Drought and fear of dehydration may turn individuals who would otherwise show each other benevolence to self-serving worriers. Boniface, a boy residing in an orphanage in a drought-ridden village, comes up with an excellent idea to save his community from water deprivation and then return to their more friendly and peaceful ways.

Fire, Wildfires

While many wonderful titles provide fire safety tips and instruction, the following books focus on the results of large-scale fires and the broader scope of fire's power and destructive force. Included also are titles that address the social and environmental impact of large-scale fires.

Allen, Nancy. *First Fire: A Cherokee Folktale*. Mt. Pleasant, SC: Arbordale, 2014. (Gr. 1-3)
Once, when the world was still dark, the animals saw light in the form of fire in a tree and tried to capture it. In trying to capture the fire, each animal was permanently changed in this pourquoi tale. It provides an excellent opportunity for discussing the power of fire to alter lives.

Bromley, Anne C. *The Lunch Thief.* Thomaston, ME: Tilbury House, 2010. (Gr. 1-5)

When Rafael witnesses the new student Kevin stealing lunches, Rafael wisely seeks answers rather than jumping to conclusions. He discovers that Kevin lives with his family in temporary housing in a rundown hotel because they have lost their home in a California wildfire. He offers friendship and food to his temporarily homeless classmate.

*Bunting, Eve. *Smoky Night.* New York: Houghton Mifflin, 1999. (Gr. 2-5)

This Caldecott-award title replays a fictional event following the Los Angeles riots' devastating fires devastating fires. In the shelter where they have been forced to stay temporarily, the young narrator's cat is missing. His cat is soon found and has been befriended by the rival cat owned by Mrs. Kim, owner of a grocery store looted in the riots. This is a true redemption story as the human characters follow their cats' example and overcome differences during difficult times.

Ehlert, Lois. *Cuckoo: A Mexican Folktale.* New York: Houghton Mifflin, 2000. (Gr. K-3)

Cuckoo is a beautiful but lazy bird that refuses to do his jobs. But when a field fire threatens to destroy the annual crops, Cuckoo must risk her vanity to save them. Ehlert's retelling of an original Mexican folktale is rich and colorfully illustrated.

Elya, Susan Middleton. *Fire Fuego! Brave Bomberos.* New York: Bloomsbury, 2012. (Gr. P-1)

The author incorporates significant Spanish words related to fire in this engaging read-aloud. The importance of fire safety and the dangers and difficulties of a firefighter's job are brought to life in this lively suspenseful tribute. It is included here because it skillfully bridges Spanish-speaking students and English-only teachers on this important topic.

Hoberman, Mary Ann. *Mrs. O'Leary's Cow.* Boston: Little Brown, 2007. (Gr. P-2)

When Mrs. O'Leary forgets to take her lantern when she leaves the barn, the cows kick it over and start a fire. The cows are saved by local firefighters in this light fiction story.

Mortensen, Denise. *Ohio Thunder.* New York: Houghton Mifflin, 2006. (Gr. K-3)

Midwest summer storms can be stunningly beautiful and refreshingly quenching but also frighteningly dangerous. An Ohio farm family quickly prepares for a menacing summer storm that brings rain needed to water crops and soak the dry land.

Nelson, S. D. *The Star People: A Lakota Story.* New York: Harry Abrams, 2003. (Gr. K-3)

Young Wolf and Sister Girl are so engrossed in gazing at the night sky that they wander too far from their village. A terrifying prairie fire leaves them stranded in the woods alone at night. After surviving the prairie fire, they must find their way back and are guided home by the spirit of their grandmother among the stars.

*Polacco, Patricia. *Tikvah Means Hope.* New York: Doubleday, 1996. (Gr. 1-5)

When California wildfires threaten homes and lives, Duane and Justine and their neighbors the Roths are forced to evacuate, and their temporary shelter will not allow animals. Thus, they must abandon their cat. The Roths have built a sukkah, a temporary structure constructed to celebrate Sukkot, in their backyard. When they return after the fire, they find everything is destroyed except the sukkah, and out of the ashes they find Tikvah, their cat.

Reynolds, Marilyn. *The Prairie Fire.* Custer, WA: Orca, 1999. (Gr. 1-3)

A young boy experiences the frightening swiftness of a windswept plain's fire's spread in this historic homesteading family story. Share this story to discuss how fighting a fire on the isolated and often desolate plains is different from in cities with fire departments and firefighters.

Floods

Creation stories, animal fiction, and folk literature abound with stories with flood themes. These titles offer great opportunities to discuss the impact of flood on all living things.

*Hausman, Gerald. *The Otter, the Spotted Frog, and the Great Flood: A Creek Indian Story.* Bloomington, IN: Wisdom Tales, 2013. (Gr. 1-3)

This powerful creation story provides a gentle reminder to prepare for impending weather and to listen to the wisdom of those who know. Spotted Frog warns of the great flood, but only Listener, the Otter, is cautious and wise enough to listen, and he builds a raft upon which to survive the flood. When the other animals are destroyed in the flood, Otter makes a new life for himself, and he and his mate become the first man and woman.

Kinsey-Warnock, N. *Nora's Ark.* New York: HarperCollins, 2005. (Gr. 1-4)

This heartwarming rural story tells of community and families helping each other during the Vermont Flood of 1927. Nora's grandpa is building a new home for Grandma which, when the river rises rapidly from excessive rain, is the only site high enough for all the surviving animals and neighbors to seek shelter. And they do: cows, chickens, ducks, pigs and neighboring residents. But Grandpa has gone out to help rescue the Guthrie's heifer and has not returned. The next day when neighbors leave, Grandma and Nora row to find Grandpa stuck in a tree.

Krensky, Stephen. *Noah's Bark.* Minneapolis: Carolrhoda, 2010. (Gr. K-2)
In this humorous rendering of the biblical tale, Noah helps each animal come up with its own unique animal sound to both increase communication among the animals and give Noah a little peace and quiet. Its light tone, lack of didacticism, and pourquoi plot make this story appropriate for the school library, even though it is based on the biblical story.

Lyon, George Ella. *Come a Tide.* New York: Orchard, 1990. (Gr. K-2)
(see the section about it taking a village)

Van Allsburg, Chris. *Ben's Dream.* New York: Houghton Mifflin, 1982. (Gr. 1+)
If this is still in the collection, consider using the flood dream sequence as a springboard for writer's response to the impact of a flood, particularly because Ben's dream is a response to rain.

Villa, Alvaro. *Flood.* Minneapolis: Picture Window, 2013. (Gr. K+)
Although this is a wordless picture book, its content is definitely not just for young readers. In pictures both startling and beautiful, Villa depicts a family's struggle to prepare for an impending storm and presents the intense themes of realization of the storm's severity, associated abandonment, and the storm's destructive force on the home. Illustrations depict a stark reality, including sandbagging and the home's internal destruction. After the family returns, they rebuild, and the story ends in hope and renewal.

Hurricanes, Hurricane Katrina, Hurricane Sandy

Publishers have done a wonderful service in honoring victims of Hurricane Katrina and other devastating storms by publishing many quality fiction and nonfiction titles. Included here are general stories about hurricanes as well as survival stories of Hurricanes Sandy and Katrina.

Barasch, Lynne. *Radio Rescue.* New York: Farrar, Straus & Giroux, 2000. (Gr. 2-4)
This unique one-of-a-kind biography tells the true story of the author's father who in 1923, as a 10-year-old child, helped save a Florida family during a hurricane with the expert use of his personal amateur radio and self-taught Morse code.

Berne, Jennifer. *Calvin Can't Fly.* Toronto, ON: Sterling, 2010. (Gr. P-2)
Calvin is an unusual bird. Instead of learning to fly like the rest of his family, Calvin spends his time with his beak buried in a book. When it comes time to fly south for the winter, however, Calvin is in trouble, as he has not yet learned to fly. But when the stormy weather turns to hurricane warnings, Calvin comes to the rescue as his reading has prepared him well for the storm.

*Hoog, Mark. *Letters from Katrina: Stories of Hope and Inspiration.* Fort Collins, CO: Growing Field Books, 2007. (Gr. 2-4)
The inspiration for this book came from a classroom in Colorado in which students were asked to write a letter of hope to accompany a book donation to victims of Hurricane Katrina. Letters were selected from children in Mississippi revealing, with heartbreaking honesty, their experiences during the hurricane. An excellent choice to use as inspiration for similar projects following a disaster.

*Larson, Kirby. *Two Bobbies: A True Story of Hurricane Katrina, Friendship, and Survival.* New York: Walker and Co., 2008. (Gr. K-3)
Although the gentle simple language deemphasizes the tragedies and stark realities of Hurricane Katrina, the realistic illustrations reveal more in this "feel-good" true story of a dog and a cat who befriend each other following the storm. After they are rescued, shelter workers discover that Bob Cat is blind and Bobbi the dog has become her seeing-eye dog.

London, Jonathan. *Hurricane!* New York: HarperCollins, 1998. (Gr. 1-3)
Two brothers swimming and snorkeling on their Puerto Rican beach discover the sky suddenly darkening. They rush home just as the family dashes to a shelter, where families crowd together to wait out the storm. The urgency of the impending hurricane is realistically shown in words and pictures.

Lyon, George Ella. *One Lucky Girl*. New York: DK Publishing, 2000. (Gr. P-3)
When the sky quickly darkens and the air stands frighteningly still in the trailer park where Nick "Hawkeye" lives with his parents and baby sister, they suddenly realize a tornado is approaching. After the tornado picks up their trailer, with baby Becky still inside, Nick miraculously finds her unharmed.

Paterson, Diane. *Hurricane Wolf*. Morton Grove, IL: Albert Whitman & Co., 2006. (Gr. P-3)
A young boy works with his parents to prepare his home for Hurricane Anna, including sketching on his bedroom window, "Go away, hurricane wolf." After the hurricane subsides, the next morning the boy's family joins other community members for cleanup. This is an excellent title for introducing very young children to hurricane preparation: boarding up windows, putting valuables in plastic, and gathering and storing batteries and water.

Sklansky. Amy. *The Duck Who Played the Kazoo*. New York: Houghton Mifflin, 2008. (Gr. P-1)
A musical duck that entertains his friends by playing his kazoo suddenly finds himself alone after a hurricane, all his friends swept away. So duck flies off to find another water home, a lake where he makes new friends. Only slightly a hurricane story.

*Uhlberg, Myron. *A Storm Called Katrina*. Atlanta: Peachtree, 2011. (Gr. K-4)
This starkly realistic story of hope in the midst of Hurricane Katrina finds 10-year-old Louis suddenly leaving his home with only his beloved horn. The family travels the debris-strewn flooded streets and arrive at the Superdome. When Louis and his mother get separated from his father, Louis uses his ingenuity and his horn to reunite them.

Wallner, Alexandra. *Sergio and the Hurricane*. New York: Henry Holt, 2006. (Gr. P-2)
Too young to remember the devastation of previous hurricanes, Sergio is at first excited at the prospect of the impending storm. His family and their community in Puerto Rico work to prepare for the storm, and after it is over Sergio has a greater respect for the power of hurricanes.

Watson, Renee. *A Place where Hurricanes Happen*. New York: Random House, 2010. (Gr. 1-3)
Four young friends in New Orleans share, in each of their alternating points of view, their joyful experiences prior to and ravaged experiences during and following Hurricane Katrina. The language is street verse, authentic, and local. The four friends have diverse, yet commonly devastating, experiences.

*Wiesner, David. *Hurricane*. New York: Houghton Mifflin Harcourt, 1992. (Gr. P-2)
Two boys help their parents prepare for an impending storm and then spend the day indoors waiting out the storm. The next day they make use of the tree in their neighbor's yard that has fallen during the hurricane, acting out imaginary scenes of pirates and jungles. Beautifully illustrated, this story can provide diversionary discussion topics during a weather front.

Power Outages

Power outages are a universal concern during disasters. Here several stories focus on the opportunity a power outage provides to bring communities together "in the dark." These stories also provide solace during outages.

Larsen, Andrew. *In the Tree House*. Tonawanda, New York: Kids Can Press, 2013. (Gr. P-2)
The power outage is simply a backdrop for this powerful sibling story of two brothers who, during the power outage, spend a summer evening in the tree house they built together the previous summer. When the brothers built the tree house they were closer, playing together constantly, but by the next summer the older brother has moved on to new interests. During the blackout the brothers spend an afternoon reconnecting.

*Rocco, John. *Blackout*. New York: Disney/Hyperion Books, 2012. (Gr. P+) (Caldecott Honor 2012)
A power outage suddenly brings the whole family together with no phone, no computer, and no appliances. As they gather on the roof to cool off, they suddenly witness natural phenomenon previously overlooked in the starry sky overhead and with a new sense of community, neighbors gather in the streets and celebrate the night together. The illustrations are lush and deep-violet hued; older readers may enjoy the almost graphic novel imagery.

Stolz, Mary. *Storm in the Night*. New York: HarperCollins, 1991. (Gr. K-2)
During a power outage a young boy and his grandfather face the storm together, and Thomas's fear of thunderstorms is quickly overshadowed by his wandering thoughts and his grandfather's story. Gorgeous deep blue illustrations truly capture the awe, the beauty, and the power of the raging storm.

Rain/Rainstorm

Rainstorms are universal and may cause great devastation or bring welcome relief. The variety in these titles provides diversions in the form of rain's rhythmic, lulling, or soothing qualities and puddle jumping to the relief from drought it brings. Of course, rain is also precursor to many disasters. Here several stories focus on the positive impact of rain.

*Bluemie, Elizabeth. *Tap Tap Boom Boom*. New York: Candlewick, 2014. (Gr. P-2)
Bluemie taps into all the senses as a sudden city rainstorm colorfully and "noisily" displays the thunderstorm's appeal. Sudden saturation makes dogs shake and umbrellas twirl, while others take cover underground. The rhythm of the rain can be overwhelmingly sensed, and the youngest readers will tap tap to the raindrops as this rhythmic urban display unfolds.

Boswell, Addie. *The Rain Stomper*. Tarrytown, New York: Marshall Cavendish, 2008. (Gr. K-2)
(see the section about general weather)

Carlson, Nancy. *What If It Never Stops Raining?* Minneapolis: Lerner, 2012. (Gr. P-2)
A mother responds to all the worries that a continual rainstorm (and other disasters) presents in this simple story time tale. This title is excellent for use with young readers who express multiple fears—unfounded and otherwise—that confront young children daily, that is, the "what-if's" in their lives.

Conway, David. *Lila and the Secret of Rain*. London: Frances Lincoln, 2008. (Gr. 1-3)
(see the section on drought)

Cotten, Cynthia. *Rain Play*. New York: Henry Holt, 2008. (Gr. P-2)
At the playground all is fun and games in the splashiness of a rainstorm until the lightning and thunder come and scare the children away, but at home they dry-up and wait for the rain to cease. The text is brief, but the visuals clearly depict the range of emotions a rainstorm can bring. The strength in selecting this title lies in the visual appeal for broader readership, as all are older African American boys.

De Roo Elena. *Rain Train*. New York: Candlewick, 2012. (Gr. P-1)
This gentle rain story follows a train as it chugs through the night during a soothing rainstorm. The rhythmic language provides train sounds against the background of rain sounds as it lulls the pajama-clad children to sleep. The gentleness of a rain storm makes this a fine read-aloud.

*Hesse, Karen. *Come on, Rain!* New York: Scholastic, 1999. (Gr. K-3)
(see the section on drought)

Lewison, Wendy Cheyette. *Raindrop, Plop!* New York: Penguin, 2004. (Gr. P-1)
This simple counting book shares the brief story of a little girl and her puppy enjoying a rainy walk. The illustrations denote the lushness of flora and fauna that the rain brings.

*Locker, Thomas. *Water Dance*. New York: Houghton Mifflin, 2001. (Gr.1-3)
(see the section on weather)

Martin, Bill. *Listen to the Rain*. New York: Henry Holt, 1988. (Gr. P-1)
This classic concept book shares the many personalities of rain in poetic onomatopoeia. Each double-page spread is illustrated with one concept and a single phrase describing that concept: a seashell, flowers, fish, a cricket, and finally a rainbow. A single concept, the rain, is viewed in all its forms.

*Rumford, James. *Rain School*. New York: Houghton Mifflin, 2010. (Gr. 1-3)
(see the section about it taking a village)

Shannon, David. *The Rain Came Down*. New York: Blue Sky Press, 2000. (Gr. K-2)
Shannon perfectly captures the dour-drenching mood that can saturate a community when the sky turns dark and the rain comes down. When the sky turns black the volume increases from animals frightened by the sudden deluge. Then accidents happen that create misunderstandings and more grumpiness until, in the final pages, the sun comes out and all is forgiven.

Shulevitz, Uri. *Rain Rain Rivers*. New York: Square Fish, 2006. (Gr. K-2)
This somewhat musical celebration of rain carries the reader throughout the community—both the city and the country—showing the impact of rain on everything as it rains and the prospect of what the rain will bring. This is an award-winning reissue of a classic tribute to a rainy day.

Spier, Peter. *Peter Spier's Rain.* New York: Yearling, 1997. (Gr. P-2)

This charming invitation reflects upon the joys found from a rainy day. Spier's child's eyes view of all the experiences, from gutters and puddles to cozy cocoa inside after a rainstorm. This wordless picture book is one that can be looked at over and over with newfound rain discoveries made with each viewing.

Stojic, Manya. *Rain.* New York: Random House, 2009. (Gr. P-2)

(see the section on drought)

Wisniewski, David. *Rain Player.* New York: Houghton Mifflin, 1995. (Gr. 1-3)

This traditional tale shares the story of a boy's challenge with and defeat of the rain god Chac in a game of pok-a-tok. The winner of two out of three of the games will determine the fate of the land, and, as readers suspect, Pik's winning ends the drought that has devastated the land.

Rainstorm: Monsoon

Krishnaswami, Uma. *Monsoon.* New York: Farrar, Strauss & Giroux, 2003. (Gr. K-7)

The beauty and tension created by an impending monsoon in India is portrayed through the eyes of a young girl as the gathering storm brings relief from the heat. The text, both descriptively and illustratively, is rich with the culture of northern India.

Sheth, Kashmira *Monsoon Afternoon.* Atlanta: Peachtree, 2008. (Gr. K-3)

A young boy in India is restless and wants family members to play with him, but there's a monsoon shower outside. Finally his grandfather agrees, and they build paper boats together, but when the rain subsides they take to the streets refreshed by the storm, and Grandfather reminisces about his own childhood monsoon experiences.

Snowstorm

Snowstorms, so often welcomed by children, can also bring a range of difficulties that children must endure or overcome. Here stories range from delightful depictions of freshly fallen snow to the characters facing raging snow storm.

Applegate, Katherine. *Buffalo Storm.* New York: Houghton Mifflin, 2014. (Gr. K-3)

This historic Oregon settler picture book depicts a young girl, Hallie, whose bravery does not extend to the impending severe snowstorm. Hallie and her beloved grandmother are both afraid of storms, but Hallie is able to confront her fear when a baby buffalo needs her assistance. When looking for a title to spark a discussion on overcoming personal fears and the pain of leaving loved ones behind, Newbery-winner Applegate's may be a good choice.

Gammell, Stephen. *Is That You, Winter?* New York: Houghton Mifflin, 2000. (Gr. P-2)

Crotchety Old Man Winter heads out in his rickety old truck in a disgruntled mood to spread snow and ice around the land in this askance-humored tale. Gammell's vivid winter pallet and splattery artist style beautifully capture the chill of this wintery tale.

Keats, Ezra Jack. *The Snowy Day.* New York: Puffin, 1963. (Gr. P-2) (Caldecott Award)

Peter wakes in excited anticipation of the first snow. He ventures out to make snow angels, a snowman, and his own tracks in the snow. This gentle classic story shares the anticipation and joy at the first snowfall and is a wonderful story for sharing the joy of first snow.

Spinelli, Eileen. *Coming through the Blizzard.* New York: Simon & Schuster, 1999. (Gr. P-2)

This quaint old-fashioned story tells of a Christmas Eve when blizzards threaten to discourage community members from gathering to celebrate the holiday. A poignant ending brings a ragged stranger who is welcomed by all.

Steig, William. *Brave Irene.* New York: MacMillan, 2013. (Gr. K-2)

When her dressmaker mother falls ill, Irene volunteers to face a treacherous snowstorm in order to get to the palace in time to deliver the duchess's ball gown. Irene faces all of the severe storm weather-related challenges along the way.

Wright, Betty Ren. *The Blizzard.* New York: Holiday House, 2005. (Gr. 1-4)

(see the section about separation)

Thunderstorm, Storm

Children are often frightened by the flash of lightning and the loud, suddenness of thunder. Sharing stories that either lighten the mood or show resolution to such storms can assuage children's fears.

Baker, Keith. *Magic Fan*. New York: Houghton Mifflin, 1997. (Gr. 1-4)
(see the section about tsunamis)

Beaumont, Karen. *Move Over, Rover!* New York: Harcourt, 2006. (Gr. P-2)
Rover is in his dog house chewing on a bone when suddenly a storm with lightning and pelting rain draws animal friends to join Rover as they seek shelter from the storm. Brief poetic text and colorful illustrations provide just the right tone for this cumulative story.

Crum, Shutta. *Thunder-Boomer*! New York: Houghton Mifflin, 2009. (Gr. K-2)
Great fun onomatopoeia fills the pages of this farm family story of an impending summer thunderstorm. Father, mother, and children are languishing in the heat, hoping for relief that comes in an all-too-familiar summer storm. Chaos follows as lightning followed by hail pelt the farm, and further madcap ensues when Maizy's favorite chicken flies the coop during the storm.

Geisert, Arthur. *Thunderstorm*. New York: Enchanted Lion Books, 2013. (Gr. P-3)
This beautifully illustrated, nearly wordless, picture book follows an impending storm as it ravages a rural farming countryside. Members of the farming community as well as the livestock prepare and respond to the storm. In one scene a twister is shown scattering trees and parts of homes. As the storm subsides the farm folk begin reparation.

Heinrich, Wendy. *I Am Tama, Lucky Cat*. Atlanta: Peachtree, 2011. (Gr. 1-3)
Cat Tama lives with her devoted monk in a remote, rundown Japanese temple. When a Samurai warrior wanders near in a storm, Tama saves his life and is then rewarded, bringing relief to the temple's extreme deterioration and the monk's poverty.

Macken, Joann Early. *Waiting out the Storm*. Cambridge, MA: Candlewick, 2010. (Gr. P-2)
Lavish illustrations complement this gentle story of a mother and daughter facing a thunder and lightning storm. As the weather worsens the two go indoors and wait out the storm in the comfort of their home, and the rhyming text creating a lulling certainty. Read with your youngest audience when they need comfort during a storm.

*Polacco, Patricia. *Thunder Cake*. New York: Philomel, 1990. (Gr. K-3)
In preparation for an impending thunderstorm, a wise babushka distracts her granddaughter by having her gather ingredients that they then bake into a thunder cake. When the girl shares that she does not feel brave, Babushka points out all the brave actions taken to prepare the thunder cake.

Ray, Mary Lee. *Boom! Big, Big Thunder and One Small Dog*. New York: Disney Hyperion, 2013. (Gr. P-2)
Usually a very brave dog, Rosie is frightened of thunderstorms and when a storm threatens she hides under the rug. It is finally when the young boy who is her owner wraps his arms around her that Rosie overcomes her fear. A gentle story to which young readers who are afraid of the sound of thunder can relate.

*Seeger, Laura Vaccaro. *Walter Was Worried*. New York: Square Fish, 2006. (Gr. P-1)
Seeger's abstract art dramatically shares the universal fears young children feel during a thunder and lightning storm. Each letter of the alphabet is represented by one child's emotional response: "Priscilla is puzzled . . . Shirley is shocked." Read this story with very young readers during a storm and let each student share similar emotions.

Stolz, Mary. *Storm in the Night*. New York: HarperCollins, 1991. (Gr. K-2)
During a power outage a young boy and his grandfather face a storm together, and Thomas's fear of thunderstorms is quickly overshadowed by his wandering thoughts and his grandfather's story. Gorgeous deep blue illustrations truly capture the awe, the beauty, and the power of the raging storm.

Swenson, Jamie. *Boom! Boom! Boom*! New York: Farrar, Straus & Giroux, 2013. (Gr. P-2)
As a storm rages outside, all the little boy's pet animals pile in bed with him. After his dog, cat, and snake join in, the little boy says, "No" to his sister joining in, but she protests, and chaos ensues in this simple picture book of a thunder and lightning storm.

Turner, Ann. *Pumpkin Cat*. New York: Hyperion, 2004. (Gr. P-2)
The gentle tale tells the story of a big orange homeless cat that seeks shelter from the storm in the book drop at the local library. Two librarians rescue him, making a home in the library for Pumpkin Cat, where he settles in among the books and the library's children visitors.

White, Dianne. *Blue on Blue*. New York: Beach Lane Books, 2014. (Gr. P-2)
The brief rhythmic text is complemented by old-fashioned scratchboard illustrations that show the changing weather pattern of a summer thunder storm, gray on gray, black on black, as it washes over the home by the shore. The night sky is lit up with radiant fireflies; the morning sky still carries violet-tinged clouds. The waves beyond the house show the pattern of the storm.

Tornadoes

Some titles here provide examples of how individuals struggle to regain their lives following a tornado. Others take a less frightening approach but provide opportunities for discussion about the importance of tornado drills and preparation.

Arnold, Marsha Diane. *The Bravest of Us All*. New York: Penguin, 2000. (Gr. 1-3)
Seven siblings face the terror of a tornado and when Velma Jean, reputedly the bravest of them all, reveals her claustrophobia at seeking shelter in the tornado cellar, it is her less-brave sister, Ruby Jane, who urges her to overcome her fear for the sake of safety in this Midwest historic fictional.

Fisher, Carolyn. *A Twisted Tale*. New York: Random House, 2002. (Gr. K-3)
In a humorous twist on a familiar theme, all of the animals on Bailey Tarbell's farm are thrown helter-skelter during a tornado and, though all survive, after the dust settles, Bailey discovers that the animals and their voices are all mixed up. Bailey attempts several means to help the animals get their voices back and finally the carnival ride the Twister solves the problem.

Lester, Helen. *Batter Up Wombat*. New York: Houghton Mifflin, 2008. (Gr. 1-3)
After a losing season, "The Champs" put their hopes in their new player from Australia, Wombat (or as they mistakenly call him, Wham Bat). While Wombat may not show batting skills, he does save the team, literally, when he digs a hole into which the team members burrow and thus saves themselves from an impending tornado.

*Long, Loren. *Otis and the Tornado*. New York: Philomel, 2011. (Gr. P-2)
This follow-up story finds Otis, the farm tractor, enjoying life with his animal friends sans the frightening bull who snorts and stomps from his pen. When a tornado appears on the horizon and all the animals head for the nearest safest spot, Otis bravely rescues the bull, who then shows a change of heart. This title is a perfect primary grade story for discussion of pre- and posttornado safety precautions and tornadoes' destructive power. (See "*Otis and the Tornado*" "Tornado Preparedness, Lesson Plan")

Prigger, Mary Skillings. *Aunt Minnie and the Twister*. New York: Houghton Mifflin, 2002. (Gr. P-3)
When Aunt Minnie rings the school bell alarm warning of the impending tornado, all the children on this Kansas farm drop everything and head for the root cellar. After the storm the children discover the house has been turned around. The illustrations wonderfully complement the extreme dangers of tornadoes without being too starkly frightening, which matches the author's lighter tone.

Sloat, Teri. *Farmer Brown Goes Round and Round*. New York: DK Publishers, 2001. (Gr. K-2)
Normally Farmer Brown's farm is well tended and cared for, but after a tornado, although they landed safe and sound back at the farm, all the farm animals' roles are confused. When another tornado arrives the animals' original lives, roles, and sounds are restored.

Tsunamis

Stories here provide examples of multiple issues surrounding devastation from a tsunami.

Baker, Keith. *Magic Fan*. New York: Houghton Mifflin, 1997. (Gr. 1-4)
Just when young Yoshi, a talented builder, feels there is nothing else to build, he discovers a fan that provides answers to his questions when he opens it. He takes his cue from the magic fan on what to build next, and he builds a boat "to catch the moon," a kite "to look over the world," and finally a bridge that reaches across the village. When a tsunami crashes through the village, Yoshi's bridge provides the high ground that saves lives.

Barron, T. A. *The Day the Stones Walked*. New York: Philomel, 2007. (Gr. K-3)
Pico's father is a stone carver on Easter Island who makes beautiful stone statues he claims hold mythical powers. When Pico is caught in a tsunami, he discovers the power of the statues that provides him guidance and strength enough to survive the wave.

Bauer, Marion Dane. *A Mama for Owen*. New York: Simon & Schuster, 2007. (Gr. K-2)
This gentle version of the true story of the orphaned hippopotamus Owen who is befriended by a giant tortoise may be the most appropriate tsunami story for the very young. The golden-hued illustrations, smiling animals, and gentle language do not provide a very authentic tone for a story that begins with a tsunami, but younger readers may enjoy this story of unusual friendship.

*Hatkoff, Isabella, Craig Hatkoff, and Dr. Paula Kahumbu. *Owen and Mzee: The True Story of a Remarkable Friendship*. New York: Scholastic, 2006. (Gr. K-4)
Six-year-old Isabella Hatkoff and her father share the story of a baby hippopotamus who, following the loss of his mother in the 2004 Southeast Asian tsunami, is rescued and brought to a natural preserve in Kenya. Here the hippo befriends a 130-year-old tortoise and thus begins a remarkable friendship. This true story presents an unusual tale of animal friendship in lavish color photographs. Maps and background information help ground this delightful tale.

Kroll, Virginia. *Selvakumar Knew Better*. Walnut Creek, CA: Shen, 2009. (Gr. 1-5)
In this retelling of a true story Dinakaran's dog, Selvakumar, senses the impending tsunami long before family members do, and he barks and whimpers a warning. When the family finally notices the incoming tsunami, Dinakaran mistakenly tries to hide in his house instead of moving to higher ground. This time Selvakumar stands his ground and tugs at his owner's sleeves, pulling Dinakaran up the hill to join his family.

Winter, Jeanette. *Mama*. New York: Harcourt, 2006. (Gr. P-2)
This picture book uses only two words to tell the story of a young hippopotamus separated from his mother during the 2004 Southeast Asian tsunami. This is one version of several accounts of the true story of baby Owen, who, after his mother is killed by the tsunami, is adopted by an old tortoise.

Volcanoes and Earthquakes

Volcanoes and earthquakes produce such a wide range of problems; attempts were made here to provide a sampling from the various issues as well as historic and recent volcano and earthquake disaster stories.

Danticat, Edwidge. *Eight Days: A Story of Haiti*. New York: Scholastic, 2010. (Gr. 1-4)
This book shares a raw but tender recounting of a seven-year-old Haitian earthquake victim who was trapped under the rubble for eight days. The story tells of Junior's bravery in the days following the earthquake, when he used positive imagery to visualize various beautiful events in his life as he waited for rescuers to free him.

George, Jean Craighead. *Dear Katie: The Volcano Is a Girl*. New York: Hyperion, 1998. (Gr. 1-3)
Katie's grandmother is quite the volcano scholar while Katie tends more toward the romance of the mythology surrounding the Kilauea Volcano. The two head off to Hawaii for a geological vacation, and each learns and shares the expertise and knowledge specific to their passions. Bright volcanic-red illustrations brighten this information-packed story.

Lee, Yangsook Choi. *Earthquake!* New York: Farrar, Strauss & Giroux, 2006. (Gr. 1-4)
In this historic story a family in Chinatown is awakened by the 1906 San Francisco earthquake and must evacuate their home, moving to a temporary shelter in Golden Gate Park. From the tented community the family witnesses the more devastating destructive forces of the fires caused by the earthquake. Fascinating facts are provided in the author's note.

Paltini, Margery. *Earthquack*. New York: Simon & Schuster, 2005. (Gr. K-2)
Using the traditional tale of Henny Penny ("The sky is falling, the sky is falling!"), the eruption of an earthquake is misinterpreted by all the barnyard animals: Lucy Goosey, Chucky Ducky, and Rickie Chickie. This time it is Chucky Ducky who is earthquacking on the loose, trying to convince the rest of the farmyard animals of their doom in this humorous spinoff illustrated by Barry Moser.

*Peters, Lisa Westberg. *Earthshake*. New York: Greenwillow, 2003. (Gr. All)
Twenty-two poems provide an interesting perspective and lend unique voices to a variety of geological topics from fossils and glaciers to continental plates. Although the poems have a light-hearted tone, their content informs while entertaining young readers in this disaster gem.

Peters, Lisa Westberg. *Volcano Wakes Up*. New York: Henry Holt, 2010. (Gr. 1-4)

Illustrated by award-winning artist Steve Jenkins, this unique title presents, in poetry, five versions of a day in the life of a volcano from five different points of view. Just as people do, members of the natural world view a volcano very differently, depending on their perspective. The poems make an excellent mentor text for encouraging similar writing.

Schaefer, Lola. *An Island Grows*. New York: Greenwillow, 2006. (Gr. K-3)

A volcano is born. This concept book presents the formation of an island from an undersea volcanic eruption. The title provides centuries of change that happens to the island as it grows from its violent beginning to a lush inhabited Island.

Watson, Jesse Joshua. *Hope for Haiti*. New York: Putnam, 2010. (Gr. K-3)

This story is told from the point of view of a young soccer-playing Haitian boy who experiences the 2010 earthquake firsthand and loses everything in the devastation. Forced to live in a makeshift home, the boy and his friends try to find joy in playing soccer, though their ball is a wad of rags, until a local hero soccer player donates a ball.

Weather, in General

Some titles that share stories related to weather can be valuable to share with young children as a springboard for weather discussion.

Barrett, Joan. *Cloudy with a Chance of Meatballs*. New York: Atheneum, 1982. (Gr. K-3)

This classic fantasy title about the strange weather patterns of the town of Chewandswallow finds mashed potato snowstorms and hailing meatballs. Use this title for a light-handed introduction to weather.

Boswell, Addie. *The Rain Stomper*. Tarrytown, New York: Marshall Cavendish, 2008. (Gr. K-2)

In this ode to going forth despite the weather, young baton twirler Jazmin is ready for the big parade. She does not allow thunder, lightning, or the torrential downpour to dampen her spirits. An excellent read-aloud with onomatopoeic weather phrases and dances in the rain that invite beginning readers to join in.

Harper, Jamie. *Miss Mingo Weathers the Storm*. Cambridge, MA: Candlewick, 2012. (Gr. P-2)

The animals visit the local weather station on a field trip and on the way there come up with many fun ways to deal with the inclement weather.

*Locker, Thomas. *Water Dance*. New York: Houghton Mifflin, 2001. (Gr. 1-5)

In gorgeous traditional oil paintings and free verse, Locker presents the many personalities of water in the earth's water cycle. From the familiar bodies of water—streams, rivers, seas—to less familiar forms such as mist, rainbow thunderhead, and storm front. The book's format invites interaction as each page's description is followed by the statement, "I am —".

Polacco, Patricia. *Meteor!* New York: G. P. Putnam's Sons, 1987. (Gr. 1-4)

One summer during her childhood, Patricia experienced the meteorological phenomenon of a meteor landing on Grampa and Gramma's farm. The next morning word spread quickly, and all the neighbors in Union City came to witness the phenomenon, creating a mass spontaneous celebration. While this is an older title, it is a one-of-a-kind story of a meteor's landing.

BOOKS THAT HEAL: STORIES TO READ FOLLOWING DISASTER

This section, divided into subcategories, provides picture books to share or read aloud following a disaster. Several subsections address some of the fallout—issues a child may have to deal with—following disasters and crises. Subsections also provide suggested titles to read to children when they need comfort or inspiration for resilience. Included also is a section of titles on fire safety and preparedness.

Bravery, Children in Literature, Children as Role Models

The picture books listed here provide a broad range of stories, the primary theme of which is noting children's bravery during difficult times. Many of these titles provide excellent springboards for discussions concerning facing fears or mentor texts for writing.

Baker, Keith. *Magic Fan*. New York: Houghton Mifflin, 1997. (Gr. 1-4)
(see the section about tsunamis)

Berger, Carin. *The Little Yellow Leaf.* New York: HarperCollins, 2008. (Gr. P-2)
A single leaf clings to the branch, refusing to fall and claiming, "I'm not ready." The story presents an excellent model for discussing the fear of the unknown and taking chances.

Bulla, Clyde Robert. *The Poppy Seed*. New York: Puffin, 1994. (Gr. 1-4)
To plant poppy seeds, Pablo bravely faces the wrath of the bad-tempered wealthy landowner who hoards water from the thirsty villagers. When the poppy seeds bloom, their beauty melts the heart of the landowner, and he shares his water with the villagers.

Butts, Ed. *S.O.S. Stories of Survival.* Toronto: Tundra Books, 2007. (Gr. 4-9)
Each of the 15 events in this collection of the world's greatest disasters shares the perspective of one survivor, several are told from or include a child's viewpoint. Historic events such as the Johnstown Flood and the fire at the Triangle Shirtwaist Factory raise important humanitarian issues. The 2004 Southeast Asian tsunami and Hurricane Katrina are personalized by a youngster's compelling stories of survival.

Cocca-Leffler, Maryann. *Bravery Soup*. Morton Grove, IL: Albert Whitman, 2002. (Gr. P-2)
Carlin, the scared raccoon kit who's afraid of everything, is sent by Big Bear on a perilous journey to gather ingredients for bravery soup. Carlin, of course, overcomes many obstacles that, upon reflection and guidance from Big Bear, serve to reassure Carlin of his own abilities and bravery.

Danticat, Edwidge. *Eight Days: A Story of Haiti*. New York: Scholastic, 2010. (Gr. 1-4)
(see the section about volcanoes and earthquakes)

Demi. *The Boy Who Painted Dragons*. New York: McElderry Books, 2007. (Gr. 1-4)
Young Ping covers the walls of his home with gorgeous paintings of fierce dragons, not revealing that he creates the dragons to overcome his own fears. The dragons he paints represent cosmic weather forces such as thunder and lightning. To face his fear Ping must meet with the Heavenly Dragon. This is a great story to help students face their own fears.

Harrison, Troon. *Courage to Fly*. Markham, ON: Red Deer Press, 2006. (Gr. K-2)
When a young girl, Meg, moves with her family from the serenity of a Caribbean Island to a frightening U.S. urban home, she must overcome many fears before settling in. Meg finds a young bird lost in a snowstorm and she, her mother, and her newly befriended neighbor nurse the bird to health. This is a great story to read to help a child overcome fears that come with relocation.

Hoffman, Mary. *Amazing Grace*. New York: Penguin, 1991. (Gr. 1-4)
Grace, usually undaunted by challenges, is told she cannot try out for the part of Peter Pan because of her gender and ethnicity. When Grace's grandmother hears this, she takes Grace to the ballet to see a famous black dancer. The dancer inspires Grace to try out for Peter Pan, she earns the part and performs it fabulously.

Kroll, Virginia. *Selvakumar Knew Better*. Walnut Creek, CA: Shen Books, 2009. (Gr. 1-5).
(see the section about tsunamis)

Markel, Michelle. *Brave Girl: Clara and the Shirtwaist Maker's Strike of 1909*. New York: HarperCollins, 2013. (K-3)
Clara Lemich immigrated to America from Ukraine to escape political tyranny only to be subjected to appalling conditions in the factory in which she works. When even men feared speaking up for union workers' rights, Clara helped lead a women's workers' walkout and subsequent strike that resulted in improved working condition. This multiple star–reviewed title clearly presents factory conditions in brief, simple language and engaging illustrations.

Polacco, Patricia. *The Butterfly*. New York: Penguin, 2009. (Gr. 2-5)
Monique lives in a French village outside Paris that is inhabited by the Nazis. One night she discovers what she thinks is a ghost sitting at the end of her bed. The ghost, Sevrine is, in fact, a Jewish girl hiding her home. The two develop a late-night friendship that ends abruptly, and Sevrine's family must flee when a neighbor spots the two girls. Based on a true story, Monique exhibits strong will and bravery daily in the ever-presence of Nazi occupation.

Hest, Amy. *When Jessie Came across the Sea*. Cambridge, MA: Candlewick, 1997. (Gr. 2-5)
This lengthy immigration story details 13-year-old Jessie's lone immigration to America from Russia, including a frightening crossing, establishing a profession as a lace maker in New York, and eventually reuniting with her beloved grandmother. Gorgeous illustrations, over-sized text, and a hardship-optimism theme make this an excellent read-aloud and discussion starter.

Krensky, Stephen. *Sisters of Scituate Light*. New York: Dutton, 2008. (Gr. 1-4)
Based on a true story, this fictionalized account relates the tale of two sisters who, during the War of 1812, outwitted British soldiers who came ashore near the lighthouse where their father was the lighthouse keeper. Father was away when the soldiers arrived, and the sisters plotted a clever means for discouraging British advances. An author's note explains the historic context.

Steig, William. *Brave Irene.* New York: MacMillan, 2013. (Gr. K-2)
(see the section about snowstorms)

Stryer, Andrea Stenn. *Kami and the Yaks.* Baltimore, MD: Bay Otter Press, 2007. (Gr. K-2)
When deaf and brave young Kami discovers his yaks are missing, he sets off in a different direction from his father's search knowing the yaks like the meadow. Kami faces the icy perils of a mountain snowstorm and soon discovers a young yak stuck in the rocks. He rushes to get help from his family. This unique story provides unusual background and details for telling the story of the child's bravery. An author's note details the lives of Mt. Everest's Sherpas.

Waber, Bernard. *Courage.* New York: Houghton Mifflin, 2002. (Gr. K-2)
Waber takes on the significant question, "What is courage?" Each page has a "Courage is" statement followed by a full-page illustration depicting a definition of courage, from simple tasting of vegetables to more universal themes that incorporate decision making and friendship. The examples address courage-making decisions that children most often must face.

Bravery, Biographies

In this section, titles are primarily biographies about people in history that have overcome obstacles related to disasters to become world renowned.

Barasch, Lynne. *Radio Rescue.* New York: Farrar, Straus & Giroux. 2000. (Gr. 2-4)
This one-of-a-kind biography tells the true story of the author's father who in 1923, as a 10-year-old child, helped save a Florida family during a hurricane with the expert use of his personal amateur radio and self-taught Morse code.

*Barron, T. A. *High as a Hawk.* New York: Philomel, 2004. (Gr. 1-3)
As a tribute to her mother who recently, suddenly died, Harriet with her father is committed to summiting the 14,000+-feet Longs Peak in Colorado. Enduring many hardships after her father abandons the summit, Harriet, who is only eight years old, becomes the youngest ever to summit this height. Details of the blizzard conditions and painful blisters are splendidly described and detailed in gorgeous illustrations. A fine testament to children striving toward a difficult but reachable goal.

Nobisso, Josephine. *John Blair and the Great Hinkley Fire.* New York: Houghton Mifflin, 2000. (Gr. 3-5)
This book tells the little-known heroic events of the 1894 Great Hinkley Fire that took the lives of over 400 Minnesota settlers. John Blair, an African American porter who worked aboard the railway train, saved hundreds of lives by helping people flee via the train, caring for them, and finally leading them to safety in a nearby swamp. Graphic realism is presented in illustrations and text.

Comfort Stories

These stories help provide comfort to children during or following a disaster or crisis.

*Harris, Robie H. *When Lions Roar.* New York: Orchard, 2013. (Gr. P-K)
This simple concept book presents several familiar scenes that frighten children and then brings resolution by depicting the first-person young narrator facing fear through personal empowerment. Illustrated by Caldecott-winner Chris Raska, this is an excellent title for sharing with very young children who are frightened by events showcased in the text such as thunder, lightning, and loud sirens.

Johnson, Nel. *Falling Raindrop.* Bel Air, CA: Tricycle, 2010. (Gr. K+)
This story personifies a raindrop's fall to earth in an impending storm. The unique perspective and unusual illustrations move this story from a simple metaphor to a potential allegory of fear of the unknown. A fine springboard for older students' writing.

Keats, Ezra Jack. *The Snowy Day.* New York: Puffin, 1963. (Gr. P-2)
(see the section about snowstorms)

LeMarche, Jim. *The Raft.* New York: HarperCollins, 2000. (Gr. P-2)
This beautifully illustrated, gentle story serves multiple purposes for young readers. A young boy is introduced to the joys of rafting when, spending the summer with his grandmother, she introduces him to the river and the sport of river rafting. The river is seen here as beautiful, peaceful, and serene.

The intergenerational story provides a wonderful backdrop for librarians seeking a story of children having to relocate with extended family.

Macken, Joann Early. *Waiting out the Storm*. Cambridge, MA: Candlewick, 2010. (Gr. P-2)
(see the section on thunderstorms)

Penn, Audrey. *The Kissing Hand*. Terre Haute, IN: Tanglewood, 2006. (Gr. P- 2)
This recent classic provides a gentle reminder of a parent's unconditional love. Chester is afraid to go to school, so his mother shares a secret: She kisses the palm of his hand and reminds him that anytime he feels lonely he should press his hand to his cheek and feel her kiss.

Rosenthal, Amy. *Plant a Kiss*. New York: Harper Collins, 2011. (Gr. P-3)
Use this story to open a conversation about the ripple effect of kindness and good will. A young girl plants a kiss in the ground and after watering and tending it, decides it's time to harvest and share. She generously spreads her kisses around; when they are depleted she returns home only to find even more have grown.

Yolen, Jane. *Owl Moon*. New York: Penguin, 1987. (Gr. K-3)
This Caldecott-winning title portrays a young child and her father taking a moonlit walk in a snow-covered woods appreciating nature and looking for the Great Horned Owl. The father calls out to the owl and they listen for the owl's reply and watch for signs. The stillness of the woods and the awe of the owl's reply make this one of the most loved of all comfort stories.

Coping with Adversities

Children who experience disasters soon learn that overcoming their fears and coping with adversity related to the event are a major part of recovery efforts. Reading stories of others who have found ways to overcome these adversities and fears helps promote healing.

*Empson, Jo. *Rabbityness*. London: Child's Play International, 2012. (Gr. P-5)
Rabbit likes to do rabbity things such as eating and burrowing, but also unrabbity things such as paint art and make music. Rabbit disappears one day, leaving nothing but a deep dark hole. The other rabbits are sad until they discover joy from the art and music the rabbit has left behind. The illustrations—which alternate between stark black-and-white and popping color—contribute to the story's powerful message and tone. Particularly interesting is Rabbit's unexplained disappearance. Discussions galore.

*Moundlic, Charlotte. *The Scar*. Cambridge, MA: Candlewick, 2011. (Gr. K-4)
The young unnamed main character in this story wakes to find that his mother has died in the night. He experiences a full range of emotions, from anger and panic to isolation and bartering, but mostly fear. He fears he will forget her; he fears his father will not be able to care for him. When he falls and scars his knee, he wishes the wound would last forever so that he could hear his mother's consoling words.

Varley, Susan. *Badger's Parting Gifts*. New York: HarperCollins, 1999. (Gr. P-3)
This animal tale presents the story of the death of Badger and the friends he left behind. Badger's friends are sad at his parting but find ways to celebrate his life. This classic has been used to help children coping with the loss of a pet.

Viorst, Judith. *Alexander and the Terrible, Horrible, No Good, Very Bad Day*. New York: Atheneum, 1989. (Gr. K-3)
While this classic Viorst story carries a light tone, teachers and students who are having difficulty dealing with day-to-day worries can find escape and humor in this gentle tale. Alexander has one silly but unfortunate happenstance after another as he goes through his day. He ends the day by hoping tomorrow will be better.

Overcoming Fear, Unfounded Fear

Battersby, Katherine. *Brave Squish Rabbit*. New York: Viking, 2012. (Gr. P-1)
Squish Rabbit is afraid of many things, including storms. As he goes to visit his squirrel friend Twitch and finds her missing, Rabbit realizes he must face his fears of both the night and storms to ensure Twitch's safety. Battersby's expression of fear is spot on regarding how the very young feel about such fears. A great read-aloud for very young readers.

Cutler, Jane. *The Cello of Mr. O*. New York: Dutton, 1999. (Gr. 2-4)
In this book that begins, "I'm afraid nearly all the time" where "Nothing is as it was" and illustrations show explosions outside her window, a young girl shares her raw, very real wartime fears and everyday

struggles. After bombs destroy the relief truck that brings needed supplies, Mr. O, a grumpy old neighbor, plays his cello in the square, and the community finds solace in his music.

McCully, Emily Arnold. *Mirette on the High Wire.* New York: Putnam Sons, 1992. (Gr. 1-5) (Caldecott)
Mirette works hard in her mother's Paris boardinghouse, which is home to many street performers. When Mirette witnesses the Great Bellini walking a tightrope, she is inspired to learn the art. When Bellini sees her diligently practicing, he helps her learn while she inspires him to overcome his fear.

Napoli, Donna Jo. *Albert.* New York: Harcourt, 2001. (Gr. 1-3)
Albert suffers from agoraphobia and spends his life in his small apartment, fearing whatever potential dangers might befall him. One day as he sticks his hand through the bars covering his window, a bird begins to build a nest in his hand. When eggs hatch and fledglings leave the nest, Albert finds the courage to go outside as well.

Reynolds, Phyllis Naylor. *King of the Playground.* New York: Aladdin, 1994. (Gr. K-2)
After Kevin decides he no longer wants to visit his favorite playground for fear of being bullied by Sammie, self-proclaimed King of the Playground, his father helps him both overcome his fear and face the playground bully who then becomes Kevin's playmate.

Snicket, Lemony. *The Dark.* Boston: Little Brown, 2013. (Gr. K-2)
This title provides an excellent metaphor for addressing children's fear of the dark. Here "the dark" is a character who lurks in the house, usually in the basement, but one night comes into his room. Lazlo musters his courage to face the dark and his fear.

Snyder, Zilpha Keatley. *The Changing Maze.* New York: Simon and Schuster, 1985. (Gr. 2-4)
Hugh lives with his grandmother and a lamb as his only companions. Granny's cautionary tale of the Ragged Hills warns what can befall citizens who succumb to greed as they seek the gold at the center of the evil wizard's Changing Maze. When the young lamb disappears into the maze, Hugh must overcome his fear to rescue the lamb.

Watt, Melanie. *Scaredy Squirrel.* Tonawanda, New York: Kids Can Press, 2013. (Gr. P-2)
Afraid of nearly everything outside, Scaredy Squirrel decides instead of camping to stay inside and watch camping on television in this humorous tale. After discovering that the cord for viewing needs an extension, Squirrel comes up with a complex plan for facing the challenge of the campground so that he can plug in. S quirrel overcomes his fear and enjoys the campsite's nighttime activities of roasting marshmallows and singing camp songs.

"It Takes a Village": Communitywide Response

Anyone who has suffered a major disaster has painfully come to know the importance of a community banding together during recovery. These stories address such community efforts.

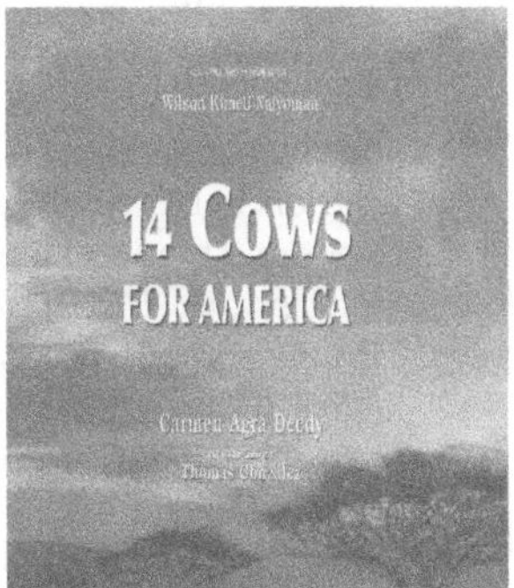

Barnett, Mac. *Extra Yarn.* New York: Harper Collins, 2012. (Gr. K-3)
Annabel comes upon a box of yarn and begin to knit. First she knits a sweater for herself and her dog, then for her classmates. It seems the more she knits, the more endless her supply of yarn is. Everywhere she creates something colorful and new, causing people's and objects' moods to lighten. When an evil archduke tries to steal her box, he soon realizes that without her generosity there is no magic.

*Deedy, Carmen Agra. *14 Cows to America.* Atlanta: Peachtree, 2009. (Gr. 1-adult)
"To the Maasai, the cow is life." When Kimeli returns from New York City to his village in Kenya, he has a story to tell his people. Once fierce warriors, the peaceful cattle herders listen intently to the story of 9/11 and ask, "What can we do for [them]?" The elders U.S. ambassadors to their village, where they perform a sacred ceremony and present the United States with 14 cows.

Lyon, George Ella. *Come a Tide.* New York: Orchard, 1990. (Gr. K-2)
A close-knit family in the mountains experiences a four-day severe weather front that brings rain, snow, wind, and floods from which they must seek higher ground. Once the storm has ended all gather to work hard in recovery and restoration efforts. This older title shows a delightful sense of community and resilience in the face of great adversity.

Rocco, John. *Blackout*. New York: Disney/Hyperion Books, 2012. (Gr. P+)
(see the section about power outages)

*Rumford, James. *Rain School*. New York: Houghton Mifflin, 2010. (Gr. 1-3)
A school story and a rain story. In this African tale a young boy in a Chadian village hurries off to his first day of school, nervous and excited. From the first day when the children build the school and through the school year, the appreciation of hard work and education pays off. As the school year closes the rainy season begins, and the school itself is washed away, only to be rebuilt again the following September. A wonderful reminder of the "It takes a village" theme that disasters often bring forth.

Villa, Alvaro. *Flood*. Minneapolis: Picture Window, 2013. (Gr. K+)
Although this is a wordless picture book, its content is definitely not just for young readers. In pictures both startling and beautiful, Villa presents a family's struggle to prepare for an impending storm, realization of severity, abandonment that follows, and the storm's destructive force on the home. Illustrations show a stark reality, including sandbagging and the home's internal destruction. After the family returns they rebuild, and the story thus ends in hope and renewal.

Loneliness, Isolation, and Overcoming Obstacles

Bunting, Eve. *Dandelion*. New York: Houghton Mifflin, 2001. (Gr. 1-4)
The loneliness and isolation of settlers' lives on the Great Plains is shared through the eyes of a young girl, Zoe, as she and her family adjust to their new Nebraska surroundings. Though historic, the girl's voice is clearly personal and so may prompt discussion of similar feelings.

Button, Lana. *Willow's Whisper*. New York: Kids Can Press, 2010. (Gr. P-2)
Willow is so soft-spoken that no one notices her, and no one can hear her. Because of this she feels lonely, isolated, and "unheard" and is often misinterpreted by her classmates. When Willow fashions a microphone from items pulled from the recycle bin, she suddenly "finds her voice," but the class bully destroys it. Willow, however, cleverly regains her confidence and voice.

Deacon, Alexis. *Beegu*. New York: Farrar, Straus & Giroux, 2003. (Gr. P-1)
Beegu is an adorable alien who is lost on earth after her station crashed. Beegu wanders to various locales where inhospitable adults turned her away. Finally, she finds a playground where children befriend her and give her a more positive impression of her visit to earth, which she tells her parents about when they arrive to take her home.

Ludwig, Trudy. *The Invisible Boy*. New York: Knopf, 2013. (Gr. K-2)
A perfect picture book to read to a class of primary students to demonstrate the importance of including everyone. Brian is a quiet "invisible" boy who no one notices; even the teacher has little time to recognize Brian's talents as she focuses on noisier children. Illustrations present Brian in soft black-and-white while noisier children are depicted in bright vivid colors. When Justin, a new student, joins the class he notices Brian, recognizes his artistic talent, and helps Brian develop his talents while encouraging classmates to include him.

McKissack, Patricia C. *Goin' Someplace Special*. New York: Atheneum, 2001. (Gr. 1-3)
McKissack share this personal story of a young African American girl who has to endure personal indignities and overcome many obstacles along the way when her grandmother permits her to get about town for the first time on her own. To get to her destination, that which the young protagonist's mother terms the "door to freedom," her local library, 'Tricia Ann experiences closed doors and racist comments.

Williams, Linda E. *The Can Man*. New York: Lee & Low, 2010. (Gr. 1-5)
Tim desperately wants a skateboard for his birthday and after watching the familiar homeless man Joe Peters collect tin cans for cash, Tim decides to fund his skateboard similarly. When Tim takes over the Can Man's territory and gathers the needed cans, he has a change of heart with what to do with the funds upon discovering Joe's need for a warmer coat. This story carries a powerful theme of caring for those in need.

Woodson, Jacqueline. *Each Kindness*. New York: Penguin, 2012. (Gr. 1-4)
This realistic fiction provides excellent inspiration for discussion on bullying, caring, and treatment of fellow classmates. Maya, a new girl whose poverty does not go unnoticed, is ridiculed and tortured by her classmates, and despite her good nature she is never accepted. After she leaves her teacher throws a rock in the water to demonstrate the ripple effect that either kindness or torture can have. This is an important story to share with classes that are enduring bullying.

Relocation

Many of these titles focus on the relocation endured by Japanese American citizens during World War II, providing an opportunity for both discussing this difficult time in American history as well as bringing another perspective to the theme of relocation.

Bunting, Eve. *So Far from the Sea*. New York: Houghton Mifflin, 2009. (Gr. 1-5)
In 1972 Laura's family revisits the California relocation camp where her father's family lived and her grandfather died during World War II. As happened to so many others, Laura's grandfather—a fisherman by trade—had his possessions taken when he was interned. This is a unique book because it is about revisiting the location of a former catastrophic event.

Herman, Charlotte. *First Rain*. Morton Grove, IL: Whitman, Albert & Co., 2010. (Gr. K-2)
A young girl is sad to leave her grandmother behind but excited for the adventure of a new home in Israel, where her family has moved. In letters and emails, the two stay in touch; Abby shares a wealth of fascinating Israeli cultural customs with her grandmother as they read each other's communiqués.

Hoff, Syd. *The Lighthouse Children*. New York: Harper Collins, 1994. (Gr. P-1)
When a lighthouse is destroyed in a storm, the lighthouse keeper, Sam, and his wife must relocate and thus leave their seagull friends, but through ingenuity they figure out a way to stay in touch.

*Lee-Tai, Amy. *A Place where Sunflowers Grow*. New York: Children's Book Press, 2006. (Gr. 1-5)
When her family is moved to an internment camp, Mai is saddened to be living where nothing grows. She becomes listless, and her parents worry about her emotional state. But when Mai begins taking an art class, planting sunflowers, and eventually making a new friend, she adjusts to her temporary housing. Told in English and Japanese.

Mochizuki, Ken. *Baseball Saved Us*. New York: Lee & Low, 1993. (Gr. 2-5)
"Shorty" is both fearful and angry in the Japanese internment camp that has been forced upon him and his family. The prisoners have formed a baseball team for exercise, entertainment, and release. The guard in the high station scares him, but Shorty focuses his anger to improve his baseball swing. After the war he is still persecuted for his race, though he continues to play baseball. A good conversation starter both about racial issues and about relocating to an unfamiliar place.

*Nogushi, Rick, and Michelle Nogushi. *Flowers for Mariko*. New York: Lee & Low, 2001. (Gr. 1-4)
After spending three years in an internment camp, Mariko and her family return home only to find their truck and possessions have been taken. They are forced to stay in temporary housing until Mariko's father can begin to rebuild his gardening business and their lives. This is a unique title in that the book *starts* with the family's release from internment and focuses on how they rebuild their lives.

Shea, Pegi. *The Whispering Cloth*. Honesdale, PA: Boyds Mill, 1996. (Gr. 2-4)
This poignant story shares the experience of living in a Thai refugee camp in wartorn Southeast Asia. A young Hmong refugee, Mai, is dislocated from her family and home by war and wishes to carry her cultural craft forward and create a "whispering cloth," a story quilt. Mai's grandmother teaches her embroidery, but to reflect and survive Mai must come up with her own personal story.

Uchida, Yoshiko. *The Bracelet*. New York: Puffin, 1996. (Gr. 2-4)
A Japanese American girl and her family are separated—her father is imprisoned—and relocated to an internment camp because of the U.S. war with Japan. While some of the story focuses on the girl's friendship bracelet that is lost in the deportation, the author clearly provides imagery related to the difficulties of relocation and forced family separation.

Reunification

Reunification is an often overlooked component of emergency preparedness. Sharing stories of reunification provides models for discussion and information sharing.

Nelson, Marilyn. *Snook Alone*. New York: Candlewick, 2010. (Gr. 1-5)
Snook, a rat terrier whose only job is to catch mice and rats, lives in isolation with the monk Abba Jacobs. When Snook and Abba are separated by a fierce storm, Snook must learn to survive on his own until,

miraculously, the two are reunited. The despair of Snook's loneliness and solitude is dramatically chronicled in poetic text and striking illustrations.

Raschka, Chris. *Daisy Gets Lost*. New York: Random House, 2013. (Gr. P-2)

Lackadaisical, Daisy the dog chases balls and squirrels in the park until she suddenly realizes she is lost and alone in the woods. Her plaintive yowl finally reunites her with her equally distraught owner.

Safety, General

Both safety in general and fire safety specifically are important topics to address. Here and in the next section are stories that have a light tone in which students can learn safe practices.

Brown, Marc. *Dinosaurs Beware! A Safety Guide*. Boston: Little Brown, 1984. (Gr. P-2)

A general safety overview told in animated illustrations from dinosaurs' personifying various safety scenarios from mild home safety concerns to outdoor safety and danger stranger. Much included to invoke class discussion in preschool or kindergarten class.

Cuyler, Margaret. *Stop, Drop, and Roll: A Book about Fire Safety*. New York: Simon Schuster, 2001. (Gr. K-3)

This is one of the few fictional stories that contains fire safety in an entertaining, nondidactic book. Jessie, who is afraid of everything, is worried about performing in front of the school the act of "Stop, Drop, and Roll" when one is on fire. The author cleverly embeds many additional fire safety directives within the story.

Moncure, Jane. *Yes, No, Little Hippo*. New York: Children's Press, 1988. (Gr. P-1)

The simple brief text shares several safety tips for very young children, including not talking to strangers, water and swimming pool safety, and bicycle safety.

Pendziwol, Jean. *Once Upon a Dragon: Stranger Safety for Kids (and Dragons)*. New York: Kids Can Press, 2006. (Gr. K-2)

In this follow-up story a young girl and her friend the fire-breathing dragon set their playground creative play in fairytale settings as they reenact several tales demonstrating through Red Riding Hood's wolf and Snow White's wicked queen the dangers of talking to strangers.

Pulver, Robin. *Axel Annie*. New York: Puffin, 2001. (Gr. K-3)

This delightful diversionary humorous story shares the antics of Axel Annie, the school bus driver who makes sure all the children get to school safely, regardless of the weather. Neither steep snowy hills nor icy windy roads can detour her from delivering Burskyville's schoolchildren safely to school.

Pulver, Robin. *Axel Annie and the Speed Grump*. New York: Dial, 2005. (Gr. K-3)

This title fills an important void, as so few titles for primary grades depict the dangers of distracted drivers and distractions while driving. Axel Annie confronts the town's worst driver, Rush Hotfoot, who is not only grumpy but a danger to all on the road as he speeds and drives while performing other tasks. A great title for opening conversations with younger children about driving safety.

Rand, Gloria. *Willie Takes a Hike*. New York: Harcourt Brace, 1996. (Gr. 1-2)

Willie is excited to explore surroundings in his new home at the junkyard, but while hiking through the nearby neighborhood, he loses his way and must use his survival training skills to stay safe.

Rathman, Peggy. *Officer Buckle and Gloria*. New York: Putnam, 1995. (Gr. K-3)

Rathman's humorous Caldecott winner is a perfect springboard for studying all types of school safety. Officer Buckle falsely assumes he is extremely entertaining in presenting safety assemblies. When he discovers that it is his dog Gloria that is entertaining the children, he retires until, of course, an accident at the school reminds him of the importance of his assembly information.

Shore, Diane. *Look Both Ways: A Cautionary Tale*. New York: Bloomsbury, 2005. (Gr. P-1)

Filbert, an energetic gray squirrel, forgets several times the importance of looking both ways before crossing the street or stepping out into traffic. Several near misses share frightening details, causing Filbert to regroup and remember important traffic safety lessons he has been taught.

Willis, Jeanne, and Tony Ross. *Cottonball Colin*. Grand Rapids, MI: Erdmann Books for Young Readers, 2008. (Gr. P-1)

Because Colin is the youngest of his nine siblings, his extremely overprotective mother decides to wrap him in cotton to keep him safe. This silly choice causes greater folly as Colin is mistaken for a large snowball and a piece of white bread. This is a fine, one-of-a-kind title with a gentle humorous approach to talking about reasonable safety precautions.

Safety, Fire

Miller, Edward. *Fireboy to the Rescue: A Fire Safety Book.* New York: Holiday House, 2010. (Gr. K-3)
In comic book–graphic novel style, this full-sized, read-aloud–friendly book shares suggestions from comic character Fireboy about how to stay safe at home (including a section on safe escape from high-rise apartments) and at school. Included also are fire's positive attributes, fire safety devices, and, of course, what to do in case of a fire.

Morpurgo, Michael. *Wombat Goes Walkabout.* New York: Candlewick, 2000. (Gr. K-2)
Wombat does not feel that he has any gifts. As he wanders through the Outback looking for his mother, the other animals brag of their personal abilities, making Wombat feel less and less confident. But when a fire breaks out Wombat saves the other animals by doing what he does best: burrowing a hole for them to hide from the fire.

Nobisso, Josephine. *John Blair and the Great Hinkley Fire.* New York: Houghton Mifflin, 2000. (Gr. 1-4)
(see the section about biographies)

Pendziwol, Jean. *No Dragons for Tea: Fire Safety for Kids.* New York: Kids Can Press, 1999. (Gr. K-3)
A humorous, light, fire safety story in which a young girl meets a fire-breathing dragon she invites home for tea. When the dragon accidentally sneezes and thus starts a fire, the young girl follows her fire safety training to escape the flames and extinguish the fire.

Separation, Dealing With

One of the most devastating issues children face following a disaster is separation: separation from home, families, friends and school. Sharing similar stories can provide children with a recognition that they are not alone in their struggles.

*Beaty, Daniel. *Knock, Knock, My Dad's Dream for Me.* Boston: Little Brown, 2013. (Gr. 1-5)
This powerfully written story reveals the range of emotions a young boy feels related to his father's absence. This unique book takes the young boy from childhood to adulthood, never really explaining why the boy and his father are separated. His deep loss is emotionally heart wrenching in both pictures and language, allowing a wide range of interpretation and readership opportunities.

Kogawa, Joy. *Naomi's Tree.* Markham, ON: Fitzhenry and Whiteside, 2011. (Gr. 1-4)
The story tells of a Japanese Canadian immigrant family who, during World War II, were condemned as "the enemy" and sent to internment. The cherry tree planted in the back yard of their Canadian home stood as a symbol, continuing to grow. The grief of losing one's home is emotionally described.

Littlewood, Karen. *When Dad Was Away.* London: Frances Lincoln, 2013. (Gr. K-3)
Although the separation in the story is because Millie's dad is in prison rather than from disaster, Millie's various emotions are universally felt by children who are separated from family or home.

Reynolds, Alison. *A Year with Marmalade.* New York: Little Simon, 2013. (Gr. K-3)
When Ella and Maddie, two best friends, must be parted for a year, Maddie asks Ellie to take care of her cat, Marmalade. It takes a while for the two to warm up to each other, but after two seasons the cat and Ellie are friends and finally when the year is over, all three are reunited. Few titles are available in which young friends are temporarily separated due to unforeseen circumstances, and this one has a bright ending, making it a great recommendation for children enduring a separation from friends or family either due to disaster or some other cause.

Wright, Betty Ren. *The Blizzard.* New York: Holiday House, 2005. (Gr. 1-4)
At first Billy is disappointed that his friends will not make it to his birthday party because of the coming blizzard. But because his home is closest to the school, he soon has more birthday friends than he imagined when the teacher explains that the roads are impassable, so students will stay overnight—not at school but instead at the nearest house, Billy's.

Chapter 11

Summary: The Library as Safe Haven

Every child needs a safe place to fall—a place where he or she can explore things without worrying about failure and judgment. A library is one of those places.

—Bill Harley, ALA

In 2014 Pew Research published findings regarding Americans' relationship with libraries. While libraries have experienced many changes over the years, one fact remains: Most people still believe in the importance of libraries providing patrons with a safe place to spend time, read, study, and research (Rainie, 2014). In 2013 Pew reported that 71 percent of parents said "a major reason libraries are important is that libraries are a safe place for children" (Miller, Zickuhr, Rainie, & Purcell, 2013). And more specific to school libraries, in his recent study on perceived school safety, Biag (2014) found that students were even more likely to see their *school* library as a familiar safe place.

A majority of participants reported the library as the safest place on campus. According to students, the library provided a peaceful venue to reflect, "to let all your thoughts out," and to "imagine anything" (Biag, 2014). Students viewed their library as "comforting," "the most organized and safest place on campus," where "fights never occur" and "where no one will harass you" (Biag, 2014).

Perhaps never in U.S. history has there been a more important time for students to believe there is one sound, secure, safe place in school where they can retreat.

"Libraries need to change from places just to get stuff to places to make stuff, do stuff, and share stuff" (Valenza & Johnson, 2009). National school library leader Doug Johnson today challenges educators to reimagine the school library space as one in which users "need to stop thinking of the library as a grocery store—a place to 'get stuff'—and start thinking of it as a kitchen—a place to 'make stuff' " (Johnson, 2013). We know "that students need a place in the school where they can learn independently; find expert help in locating and using high-quality information; practice group learning skills; and feel welcome, safe, and valued" (Johnson, 2013). Valenza and Johnson further encourage all members of the school community to view the library as a social learning space, a teaching space, a production and presentation space.

And I would add to his list, a safe space.

The fact that students already find that the library provides a "sense of calmness" "free of worries or distractions" and "safe from the loud surroundings" (Biag, 2014) demonstrates the library's potential for refuge when it is needed most: following a crisis or disaster. Biag's findings present important understandings from which to launch safe haven efforts.

If children are introduced to the concept of their school library as a safe haven *before* a disaster strikes, they will be more likely to retreat to the library for that sense of safety during or following a disaster. What else then may be needed, postdisaster, to ensure that students return to the school library as their safe haven?

The most critical characteristic that students must sense to feel that the library is their safe haven is the tone of the facility itself, and this is most likely established by the librarians. This spirit of safe haven is established through the librarians' professional disposition, through the facility, and through the library's physical environment.

PROFESSIONAL DISPOSITION: THE "WELCOME"

A library gives every citizen equal access. It's a community space. It's a place of safety, a haven from the world.

—Neil Gaiman, ALA poster

Following a crisis or disaster, even the most welcoming librarian may be challenged to "find her happy face," but it is that welcome children will crave most when they are in greatest need. Imagine children's relief in knowing there is a safe place for them to retreat, where there is a welcome smile upon entering the library.

That welcoming smile when patrons walk into the library goes a long way at the beginning of the healing process. In a school that has suffered a disaster involving the entire school, where everyone has been impacted, many children feel fractured and lost, and they will seek a warm place to land or retreat. A welcome demeanor is a priceless commodity of solace.

PERSONAL STORY: Welcoming Walk-Ins

During my career change from school librarian to library educator, it became one of my tasks to find mentors for teachers in my university's school library certification program as, during their coursework, they became trained as school librarians. As I visited many school libraries, one of the things I noticed right from the beginning was the degree to which children felt welcome in the library. It was highly evident on their faces as they walked in the door. In libraries where the librarian's professional disposition created an open and welcoming environment, students entered the library confidently—knowing where to go, what they wanted to do while they were there, and usually with a happy expression on their face(s).

In some libraries, however, students entered hesitantly, their hall pass in hand leading the way, knowing that would be the first question they were grilled on: "Where's your hall pass?" In an exchange with a less than-friendly librarian, usually the next set of questions, while the initial question may have been, "What are you here for?" carried more of a "Why are you here disturbing me?" tone.

I challenge librarians to look closely at the faces of students as they enter the library. Their expressions will tell you how welcome they feel.

—Christie Kaaland

PROFESSIONAL DISPOSITION: OUTREACH

Following a crisis or disaster may be a time librarians must actively reach out to the population of students and staff they serve. Students will not know that some of the predisaster policies that were in place have been temporarily set aside unless the librarian reaches out to students and staff and offers the library as a safe haven.

Whether it is a furniture rearrangement in a corner of the library officially labeled "Safe Haven" or some other change in venue or procedure, spend a bit of time "advertising" the library's role in healing. Go into classrooms and announce this along with any additional roles of recovery in which the library will engage.

Teachers may also need that place of refuge. Consider invitations to staff as part of your postdisaster outreach. Perhaps provide an after-school social event such as an educators' book club or healing club where the librarian gives book talks of suggested titles recommended for reading to students postdisaster; this can be a healing event as well as a valuable resource for teachers (see Chapter 10).

PROFESSIONAL DISPOSITION: SOCIAL ENGAGEMENT

Students need social interaction with peers and professionals as they experience the range of emotions following a disaster. Librarians may open areas of the library for students' social engagement, whether it is to debrief and discuss events or completely escape from the current event that has transpired in the school.

> Developing meaningful social and emotional connections with students and creating a climate of mutual respect are essential to keeping schools safe. Students felt connected to the school when they believed someone in the school need[ed] them and cared for them. Simple and genuine gestures such as regularly greeting students, talking to students, and addressing students by name, help to make students feel connected and part of the school [library]. (Pollack, 2008)

LIBRARY FACILITY

Along with the librarian's welcoming smile, promote a safe haven visually through a cheerful and inviting facility. It may mean taking down a few posters highlighting library rules or schedules and replacing them with simple signs of encouragement or invitation.

> Putting a large welcome sign on the inside of the library door and leaving the door ajar whenever possible may encourage students to poke their head in and see minor welcoming changes that encourage them to visit. Add simple displays and invitations that advertise the library as a safe haven to encourage students who may be seeking such a refuge.

If the library has outside display areas, post "healing stories" that are visible to passersby. Displaying a few helpful resources, whether professional material for teachers and parents or resources that community members (including students) will need to access can increase the library's role in recovery.

PHYSICAL ENVIRONMENT

At a time where the "library as place" is being challenged by the ubiquity of technology, while students will still be able to have their information needs met outside the physical plant, following a disaster, having access to a warm and inviting physical environment may be just as important, if not more so.

If at all physically possible, rearrange or bring in comfortable seating in one corner of the library where students feel welcome and can retreat, regardless of their reason for being there. Provide this comfortable physical environment where students can gather to comfort each other, to share concerns, to debrief, as well as to sit quietly and reflect, read, relax, or regroup.

Following a disaster, making just a few minor changes to the library can provide a sense of welcome, of security and belonging.

A SAFE HAVEN OF RESPECT AND HONOR

In simple ways, the library can stand as a model of support, honoring that culture of respect that so many emergency preparedness professionals have determined is the key ingredient for increasing the school's safety and recovering following a disaster.

As time passes, the library can return to a normal schedule, but students, having endured a crisis or disaster as the most frightening yet memorable experience of their school career, will long remember that the library provided relief as a safe haven.

Appendix A

DISASTER RISK ASSESSMENT CHART

Assess your current school and school library for the following disasters. Rate each according to the likelihood of the event occurring. Consider your geographic location. Assess the likelihood using the following scale (0–5).

Highest likelihood of occurring = 5
High = 4
Moderate = 3
Less likely = 2
Least likely = 1
No threat of occurring = 0

Disasters: Natural	Rate	Priority	Notes
Avalanche			
Blizzard			
Cyclone			
Drought			
Earthquake			
Extreme cold			
Fire			
Flash flood			
Flood			
Heat wave			
Hailstorm			
Hurricane			
Ice storm			
Landslides/mudslide			
Rainstorm (severe)			
Sandstorm			
Snowslide			
Snowstorm			
Tornado			
Tsunami			

Disasters: Natural	Rate	Priority	Notes
Volcanic eruption			
Wildfire/forest fire			
Windstorm			
OTHERS*:			

Disasters rated 5 should be given top priority for preparedness. However, other disasters may have a check (x) for priority based on criteria other than likelihood of occurring.

Use the same scale to assess the likelihood of the following human-made or biological disasters. The school district may have data on these potential threats based on recent events.

Disasters: Human-Made and Biological	Rate	Priority	Notes
Active shooter			
Bioterrorism			
Biohazard/hazardous material			
Bomb threat			
Chemical hazard/threats			
Civil disorder			
Contaminated food			
Industrial accident			
Infectious disease			
Forest fire/fire			
Mold			
Nuclear disaster			
Pandemic/epidemic			
Plumbing/water			
Power outage			
Terrorism			

What are the potential consequences from these disasters to your school and/or library?

Use results for information gathering, response priorities, book orders, adding links to your school library website, and for other key information.

*Other disasters unique to each school, community, or region may be added to this list. For example, in some communities various insect or rodent infestations may occur.

From the data above, list your most likely disasters:

	Potential for Loss of Life	**Potential for Property Damage**	**Length of Recovery/ Time Frame**	**Notes: What can you impact now?**
1.				
2.				
3.				
4.				
5.				

FACULTY-STAFF TABLETOP DRILL SCENARIOS

Scenario 1: Noncustodial Parent

During Friday after-school student dismissal the two recently divorced parents of student Samira are standing outside near the school angrily arguing about with whom Samira should be staying. It is well known that this particular mother is very volatile and angry about the divorce, and she has recently been suspected of arriving at school to pick up Samira with alcohol on her breath. It appears the scene between the two parents is quickly escalating.

Discussion:

1. As one of the in-charge supervisors for after-school dismissal, you have authority here. What would your first action be?
2. What might you worry about regarding this mom; how might she respond?
3. Who else should be involved? How?
4. What special concerns might follow?

Scenario 2: Unfamiliar Visitor

Two staff members, Sarah and Sally, notice (they see him from behind) an unfamiliar man in the halls who is walking rapidly, mumbling to himself, and wearing a large overcoat under which the staff members suspect there could be a weapon. Many students are present in the halls and are nervously moving away from the man. Sarah and Sally know that he has not seen them. No other adults are visible in the halls.

Discussion:

1. Sarah and Sally, would you approach the man in the hall?
2. What are the first actions you should take?
3. What inevitabilities might arise? How might the man respond to each?
4. What do you need to know about the setting to determine actions?

Scenario 3: Hostaged Teacher

Faculty member Joe walks past his colleague's classroom. Glancing in, he notices the teacher is being held hostage by an unfamiliar man. Students are present. It appears the man may be holding a weapon at the hostaged teacher's back. Joe is sure the hostage taker did not see him glance in the classroom window.

Discussion:

1. What do you thing Joe should do first?
2. What inevitabilities might arise? How might you respond to each?
3. What action might the principal take?
4. What special concerns might follow?

Scenario 4: Student Confrontation

Stories of off-campus confrontations between two groups of students from the school are circulating. Although no student has officially complained, teachers have overheard students voicing concerns of potential violence. It is rumored that the escalating turf challenge will take place after school at a designated nearby location. Several students have heard the fighters plan to bring guns.

Discussion:

1. What action, if any, should teachers and/or administrators take?
2. Who else both within and outside the school should be contacted? How?

Scenario 5: Known Individual from Outside: Targeted Violence

As the school day begins, no student or staff member questions the principal's wife walking through the halls. She stops by the office and cheerfully asks where her husband is. Office personnel explain that he is in the gym (the morning gathering place for all arriving bus students). She proceeds to the gym where her husband (the principal), several other teachers and staff, and over 100 students are gathered. She walks quickly toward her husband and pulls from her purse a small handgun. Aiming it at him, she exclaims, "Who is she?!"

(Note: Unbeknownst to staff members, the principal and his wife have been seeking counseling for her unfounded excessive jealousy.)

Discussion:

1. Teachers and staff in attendance, what is the first action you should take?
2. What inevitabilities might arise? How might you respond to each?
3. What action might the principal take?
4. What special concerns might follow?

Find additional scenarios at

http://rems.ed.gov/docs/repository/REMS_000048_0001.pdf
http://www.opsb.us/wp-content/uploads/2012/10/OPSB-Emergency-Procedures-Guide-.pdf

Sample Response Drill for School Stabbing (Perpetrator Is No Longer At Large)

First Adult or Teacher Responder Roles:*

- Stay with the victim
- Send someone to get the school nurse
- Ask, "Is anyone trained in first aid/CPR?"
- Administer first aid as needed/as able
- Send someone for further assistance
- Inquire about witnesses
- Asks students with cell phones to call 911
- Send for crisis team members
- Assist police in their investigation

Principal/Administrator Roles, Initial:

- Contact local police, 911
- Verify victim, personal and family information
- Notify the superintendent
- Activate the school crisis team
- Notify staff to gather information for accuracy in the report
- Meet with police and first responders
- Refer media to the superintendent
- Monitor victims' progress and report

- *Principal Secondary Roles*:
- Determine if evacuation is necessary
- Interview witnesses individually
- Compose/deliver an announcement to the school, during morning announcements
- Calls for a moment of silence, if needed
- Announce "All clear" when appropriate
- Debrief with all staff
- Complete official documentation form

Counselor Roles:

- Remove all students from the scene
- Interview witnesses individually
- Assist the crisis team
- Meet with students in distress
- Contact parents of traumatized students
- Assistant the principal with witness interviews, if needed
- Increase amount of visibility in hallways and other student gathering places

Crisis Team Member Roles:

- Remove nonvictim witnesses from the site
- Cover/remove the weapon from the site
- Keep track of the weapon used
- Establish triage center
- Locate witnesses
- Document witness names and contact
- Ask witnesses to document detailed observations of the event
- Isolate/relocate witnesses from others
- Contact families of witnesses and victims
- Designate a witness and student reunification location
- Share counseling service information
- Call an additional counseling service if needed
- Accompany the victim(s) to hospital
- Assist police in the investigation

***Whether the stabbing happens in the classroom, hallway, or other school location, the first in-charge adult on the scene must accept first adult responder responsibilities.**

Additional Roles if Perpetrator Is Still Longer At Large

Principal/Administrator Roles:

- Announce official lockdown
- Announce "All clear" when appropriate

Crisis Team Role:

- Help identify perpetrator and distribute description

School Lockdown Drill Checklist

After a school lockdown drill has been established, following a drill, this check list can help determine its effectiveness and comprehensiveness. As with all emergency planning, the lockdown drill should be a dynamic rather than static plan. After reviewing results of this checklist, make appropriate changes to the established lockdown drill.

I. Who Was Involved **in the Lockdown Drill** Planning?	Yes	No	Notes
Faculty			
Support Staff			
Students			
Parent(s)			
Community and/or Local authorities			
II. Roles and responsibilities: Who Was Involved in the implementation of the Lockdown Drill?			
All school staff adults followed procedure as trained			
Incident commander acted as leader			
Documentation of procedure included all actions			
III. What **variety** was applied to the drills?			
Variety in time of day			
Variety in school routine(s)			
Variety in scenarios imagined			
Variety in who responds/how			
Variety in type of drill			
Variety for in-charge personnel (IC)			
IV. What **communications** were established for during drills? Were the detailed? Specific?			
Incident Commander communicated lockdown clearly			
Roll Call/attendance: staff, Report in			
Roll Call/attendance: students, Report in			
Codes were clearly established			
Language simply, easily identified, consistent terms			
Everyone understood responsibilities and commands			
V. **Other considerations**			
Response time: Lockdown happened in timely manner			
Implemented plan for special needs students			
How often will this drill be held?	Monthly	Quarterly	Annually

SCENARIO: DISASTER PRIORITIES
EXERCISE INSTRUCTIONS

Please consider the following scenario and the associated disaster-related issues. **Your task is not to attempt to develop solutions for these problems.** Instead, based on your discussions and input from all people in your group, please come to consensus on ranking numerically the problems in order of their importance.

Use the Disaster Priorities Matrix Table to record your priorities.

Select a spokesperson for your group to report on your decisions in 10 minutes.

Scenario

Scene: Camelot Elementary School, K-5, 612 student count. The town's population is 18,000, and it is located 51 miles from the state capitol, whose city population is 185,000.

Thursday, January 13, approximately 11:30 a.m. An unprecedentedly severe snowstorm has hit the region. As reported by the National Weather Service, there are uncharacteristically cold temperatures with severe and gusty winds.

Fire, police, public works, and other public agencies are out in full force, acting in accordance with local protocol. They will be assessing damages and initiating an immediate response when they can. Their resources are limited by this unexpected storm, and they are trying to call in all off-duty staff. The local emergency operations center (EOC) is activated, and officials and staff are scouring communities to assess potential needs and set priorities. The state emergency operations center has been activated. The governor has been contacted to declare a state of emergency for the entire state, and media are working to warn everyone to stay indoors and consult local media for further news.

Students are in school; the weather was not predicted to be this severe and the storm hit with unexpected speed, allowing no time for release of students or for informing families. The region, including this elementary school, has potential for damage to critical infrastructure that may result in power outages and structural collapse. Some roads and bridges that give access to the school are blocked. Heavy snow and high winds have taken down power lines and spread power outages throughout the community. Temperatures are dropping in the school.

Many streets are icy and dangerous; the community has been warned to stay off roads to leave them free for emergency vehicles. Communication capabilities are severely disrupted due to some power failure; radio communication for emergency services is severely reduced. What land line and cellular phone service is still working is overloaded, and few calls can get through. Commercial broadcast capability is reduced.

At this time, this is all you know.

Disaster Priorities

Issue 1: Identification of All Students in Attendance

Some students walk distances up to a mile to get to school. The speed with which the storm hit prevented administration in the main office from ensuring an accurate attendance record from morning attendance because office managers were dealing with impending storm calls.

Issue 2: Inform Some Students of Potential Relocation upon Release

The unexpected speed and severity of the storm increases the potential for storm victims among both students and their families. Numerous inquiries are already being reported to the school regarding missing people. A few calls that have gotten through have indicated incidences of hospitalized parents of students in the school.

Issue 3: Treatment of Potential Injured Students

Currently no severe injuries have been reported; however, several students have indicated potential hypothermia from their walk to school. The only medically trained staff is the school nurse, who is shared among three schools. She is not in attendance at this school today. The requirements for treating hypothermia and potentially injured students go well beyond the capability of school personnel at this time. Though not all are impassable, roads are unsafe due to icy conditions, and there is virtually no ambulance service.

Issue 4: Informing Students of the Storm's Impact

Students are currently being held in their individual classrooms. Reactions, as reported by a parent class-to-class volunteer runner, range from younger students in tears to anxious inquiries from all grades. A uniform message needs to be composed and delivered to all classrooms.

Issue 5: Uniform Communications to All Families and the General Public

Currently phone service is overloaded and stands at about 20 percent capacity. The emergency alert system is operable. A means must be established to develop and make information and emergency instructions available to families and others.

Issue 6: The Gymnasium's Flat Roof Has the Potential for Collapse

Heavy snow is building up at a rapid pace, and school officials are concerned that the building's flat gymnasium roof may fail structurally. The gym also serves as the school's cafeteria. The roof is making frighteningly "creaky" noises. Snow needs to be removed from the roof.

Issue 7: Cell Phone Communication Decisions

Most antennas have been damaged, leaving public emergency services communications among first responders at about 20 percent capacity. In the school most staff members (and a few fifth graders) have cell phones, which will start running low on battery power soon. Battery power ranges from 5 to 10 hours depending on use, phone capacity, and previous battery level. With the numerous power outages, they will not be able to be recharged effectively.

Issue 8: Identification and Distribution of Student Health Information

Some students in the school have health issues ranging from severe asthma to milder allergies. Current known issues include two students with severe asthma and one who has insulin-dependent diabetes, though other health issues among students are rumored. Disaster kits are allergy free. Students have not yet had lunch.

Issue 9: Restoration of Commerce and Business

Early calls were received from local businesses about special assistance being needed to get stores open to provide services, from donuts to batteries. Though many requests are being received for restoration of services, damages make many businesses unsafe.

Issue 10: Plan for Extended Stay in School

Many schools have been damaged, including no power, but the children are still in school. School officials will release children only to parents or guardians, but many parents cannot reach the school. Information from the district indicates that few schools are prepared for sheltering in place. The severity, the length, and the capriciousness of the storm are as yet unknown, and it is also unknown whether all or any students will need to remain in school for an extended period.

DISASTER PRIORITIES EXERCISE MATRIX

Priority	Number	Issue
		Issue 1: Identification of All Students in Attendance
		Issue 2: Inform Some Students of Potential Relocation upon Release
		Issue 3: Treatment of Potentially Injured Students
		Issue 4 : Informing Students of the Storm's Impact
		Issue 5: Communications to Families and the General Public
		Issue 6: Gymnasium Flat Roof Potential for Collapse
		Issue 7: Emergency Phone Communications
		Issue 8: Identification And Distribution of Student Health information
		Issue 9: Restoration of Commerce and Business
		Issue 10: Plan for Extended Stay in School

Appendix B
Sample Lesson Plans

OTIS AND THE TORNADO

Lesson: Tornado Preparedness

Grade Levels: K-2

Time: 20–30 minutes

Introduction: This is the story of the friendly farm tractor Otis and his animal friends who are faced with a powerful tornado. The story provides an opportunity for young children to discuss what to do in case of a tornado.

The goal of this lesson is to give very young school-age children who have not experienced a tornado the opportunity to read about a tornado in a nonemergency setting. The story addresses preparing and responding to tornadoes, as well as witnessing a successful recovery.

Note: Of all tornado fiction selections for young readers, this book offers excellent tools for tornado discussion. The story includes details in illustrations and text on weather patterns as a tornado approaches and a realistic depiction of the tornado event. The story also offers important themes of tornado preparation, the value of communities working together, and rebuilding and recovery efforts.

Lesson Objectives:

- Learning best practices for preparing for tornadoes.
- Recognizing the importance of communities working together during disasters.

Common Core Standards:

CCSS.ELA-LITERACY.RL.1.7: Use illustrations and details in a story to describe its characters, setting, or events.

CCSS.ELA-LITERACY.RL.1.1: Ask and answer questions about key details in a text.

CCSS.ELA-LITERACY.RL.1.2: Retell stories, including key details, and demonstrate understanding of their central message or lesson.

CCSS.ELA-LITERACY.RL.1.3: Describe characters, settings, and major events in a story, using key details.

Resources:

Long, Loren. *Otis and the Tornado.* New York: Philomel, 2011.

Instructional Roles:

This lesson can be taught as a single library literature lesson or as part of a longer unit on tornadoes or disasters.

Librarian's Role: Read this story aloud to students, using the discussion questions to enrich the story and guide discussions about students' potential fears related to tornadoes.

Lesson Plan Enactment:

Prestory discussion: Begin by asking students:

- *How many of you have experienced a tornado?*
- *Do you know what to do to stay safe when a tornado strikes?*
- *What are the things that you have learned about tornadoes?*

Lesson introduction: If you have not experienced a tornado, many of you may have heard tornado warning alarms or heard family members discuss what to do in case of a tornado. In this story we are going to read and listen for examples of good choices Otis makes in preparing for the tornado.

Lesson procedure: Show the illustrations on each page as you read the story, giving children the opportunity to respond to the illustrations as well as the story line. This story is best completed in one reading. The following questions may prompt a deeper discussion:

- *When the sky began to turn dark and the wind blew, Otis felt "something different" deep down in his pipes. What was different? How did Otis know this?*
- *What did the farmers do to prepare for the tornado? How could you tell that this tornado came very suddenly?*
- *Why did Otis encourage all the animals to leave the barn to lead them to Mud Creek, to the lowest part of the farm?*
- *Do you think the bull would have been okay tucked under his shed? Why or why not?*
- *The animals stayed huddled in Mud Creek long after the tornado had passed. Why?*
- *What do you think the animals thought when they discovered all the destruction from the storm? What do you suppose they did next?*
- *Do you think the tornado in this story is realistic? Why or why not?*

Lesson conclusion:

Let students talk about their own tornado recovery ideas.

If librarians assign the extension (number 4), let students tell about their illustration.

Lesson extension:

1. Prior to reading this story roll out a long piece of butcher paper on the floor and have felt pens or other drawing materials available. Following the story, tell students the butcher paper represents the farm and the field. Invite students to draw one reconstruction activity they would do. Ask them to suggest other repairs that may not be included in the pictures. Give the following examples:

- *I would like to replant corn.*
- *I would like to fix the fence.*
- *I would like to turn the car upright.*

2. Let each child draw his or her own repair picture for the farm.

Student assessment/ student reflection:

Determine if students learned objectives by:

1. Student mural responses
2. Informal assessment of student learning from student discussion response and questions following the mural drawing
3. Individual student written and drawn story response

Professional Reflection:

Indicators of this lesson's success include:

- Students level engagement in the story and follow-up discussion
- Students share additional questions or information about tornadoes

Additional Resources:

The following titles may be used to complement the story. Alternatively, recommend students check them out if they wish to read additional books on this topic:

Nonfiction for young readers:

Chambers, Catherine. *Tornado.* Portsmouth, NH: Heinemann, 2001.

Gibbons, Gail. *Tornadoes!* New York: Holiday, 2009.

Additional fiction:

Arnold, Marsha Diane. *The Bravest of Us All.* New York: Penguin, 2000.

Fisher, Carolyn. *A Twisted Tale.* New York: Random House, 2002.

Lester, Helen. *Batter Up Wombat.* New York: Houghton Mifflin, 2008.

Osbourne, Mary Pope. *Twister on Tuesday: Magic Treehouse #23.* New York: Random House, 2001.

Prigger, Mary Skillings. *Aunt Minnie and the Twister.* New York: Houghton Mifflin, 2002.

Sloat, Teri. *Farmer Brown Goes Round and Round*. New York: DK Publishers, 2001.

EMERGENCY GLOSSARY

Name______________________________________ Date____________________

Glossary: A list at the back of a book that explains or defines difficult or unusual words and expressions used in the text.

Introduction

During an emergency it is important that everyone in the school understands the words that are used to inform or instruct actions, but sometimes words have more than one meaning. Your task is to create a glossary of common terms used in an emergency so that everyone will understand the terms used during an emergency or emergency drill.

Instructions

In this lesson students will work with a partner to define important emergency terms that are used during drills or real emergencies. You may use print and online dictionaries for help but remember the definitions you are providing should be specific to emergency preparedness.

Term	Definition
First responder	*Local fire officials and police who are prepared to respond and are called to the scene of an emergency*

EMERGENCY GLOSSARY

Lesson: Emergency Preparedness and Disaster Glossary

Grade Levels: 2–4

Time: Two days, 30 minutes each lesson, or one day for 50 minutes

Introduction:

This lesson introduces students to the use of a glossary as a way to define terms on a particular subject so as to immediately understand what is meant in an emergency when specific terms are used.

Lesson Objectives:

- Learning the use and importance of a glossary
- Identifying emergency preparedness terms
- Understanding that glossary terms are not necessarily unfamiliar words but may be words with new or specific meaning

Common Core Standards:

CCSS.ELA-LITERACY.RI.2.4: Determine the meaning of words and phrases in a text relevant to a grade 2 topic or subject area. (second grade)

CCSS.ELA-LITERACY.RF.3.4.C: Use context to confirm or self-correct word recognition and understanding, rereading as necessary. (third grade)

CCSS.ELA-LITERACY.RI.4.4: Determine the meaning of general academic and domain-specific words or phrases in a text relevant to a grade 4 topic or subject area. (fourth grade)

Resources:

Print dictionaries

Online dictionaries

Nonfiction emergency preparedness books

FEMA or other emergency website

The school or district's official emergency preparedness plan

Librarian's Instructional Role

This lesson may be taught in response to a school having experienced a disaster or a drill in which the terms that were used were not fully understood. Alternatively, this lesson can be taught as part of an emergency preparedness curriculum.

Lesson Plan Enactment:

Prelesson Introduction: Set nonfiction books containing glossaries on student tables prior to their arrival to the library.

Glossary Introduction: *On your desks are nonfiction books that contain a glossary. Look in the back of one of these books and find the glossary. From looking at the example glossary, can anyone tell me what a glossary is?* (Discuss glossary's use.) *When do you think the glossary might be particularly important?*

Lesson Introduction:

Discussion: Begin by asking students to discuss the following:

Have you ever been faced with an emergency situation where you needed to act quickly and understand what to do? One of the most important things to note for a quick response in an emergency is a common understanding of the words and phrases that are used. For example, everyone knows what to do when instructed to "Stop, drop, and cover" in an emergency because you have been taught this response.

Lesson Directions:

Distribute the school district's emergency preparedness plan, one per student.

Read through the school district's emergency preparedness plan together and as a class find 12–15 important words and phrases that are key to understanding what to do in an emergency. After agreeing on these terms, select 8–12 of the words to include in an emergency preparedness plan glossary.

Lesson Procedure (lesson idea 1):

Directions: *In this lesson, students will work with a partner to define important emergency terms and phrases used during either a drill or real emergency. Use print or online dictionaries for help but remember the definitions you are deciding on are specific to emergency preparedness.*

1. Assign student partners.
2. Distribute the Glossary Worksheet and print dictionaries, and have students log in to online dictionaries.
3. After students have completed their glossary definitions, let students regroup several times until they come to an agreement about the wording of the definitions of each term.
4. Using document camera, compose the final agreed-upon glossary.

Alternative (lesson idea 2):

Instead of having each student group create the same glossary of terms, this lesson can be taught as a natural disaster lesson, and each group can be given a different natural disaster for which to create a glossary.

Example Disaster Groups: use the following natural disaster topics to create a glossary of important words and phrases that are necessary for understanding characteristics of the disaster:

- Hurricane
- Tornado
- Earthquake
- Volcano
- Tsunami
- Flood
- Fire
- Avalanche
- Mudslide or landslide

Lesson Conclusion:

After all students have agreed on the "official definitions" of the glossary terms, type these up as an addendum to the district's emergency preparedness plan and add the glossary as a class contribution to the school's emergency preparedness plan.

Lesson Extension:

Students may wish to present this glossary to the school board as an addition to the school's emergency preparedness plan.

Student Assessment/Student Reflection:

Students' worksheet demonstrates an understanding of multiple-meaning words.

Students' worksheet demonstrates an understanding of glossaries.

Students may wish to reflect on the process for agreeing on definitions.

Professional Reflection:

Indicators of this lesson's success include:

- Students' use of glossaries for other purposes.
- Student presentations of their work to school or school board reflect interest in a call to action.

EXPLOSION, UNKNOWN

Lesson: Disaster Preparedness: *The Unexplained Explosion*

Grade Levels: 3–5

Time: 40 minutes

Introduction:

This skit offers students the opportunity to read and discuss many possible reactions to a sudden explosion that could be a potential emergency or disaster.

Lesson Objectives:

- Understanding various responses in an emergency
- Recognizing appropriate responses during potential disasters

Common Core Standards:

CCSS.ELA-LITERACY.RI.3.6: Distinguish their own point of view from that of the author of a text. (3rd grade standard)

CCSS.ELA-LITERACY.RI.5.1: Quote accurately from a text when explaining what the text says explicitly and when drawing inferences from the text. (5th)

Resources:

A script for each student

Librarian's Instructional Role:

The librarian may decide to read the role of the librarian or allow a student to play the role.

Lesson Plan Enactment:

Prestory discussion: Begin by explaining to students:

You have done fire drills and lockdown drills in class. Today we are going to read a skit. This skit is another type of drill that helps prepare for emergencies. As we read the play we'll stop and discuss some of the decisions made and events that happen.

Lesson Procedure:

1. Distribute scripts. *Assign roles* in the script to students in the class.
2. *Read* the skit aloud together. Note: As you read the skit there are several events you will want to discuss. You may decide to read the whole script and discuss after reading or stop during the reading and discuss as you come to them. Or you may wish to read through once and read through a second time, stopping to discuss the important points (see step 3 for discussion questions).
3. *Discussion questions:*
    1. "No one notices that Stoney is not in the room." from the script. *What do you think happened to Stoney?*
    2. "I am okay but I should get my inhaler just in case." *Do you think Ms. Smythe made the right choice by not getting the inhaler right away? Why or why not?*
    3. "Ms. Smythe quickly joins Petra." *Do you agree with this decision? Why or why not?*

4. "Sam, would you come over here and stay near Petra?" *Here she asks Sam to leave his place. Do you agree with her decision? Why or why not?*
5. "There are way more emergencies that are just some kind of accidents than anything you mentioned. " *One really good decision Sam made here was not to mention "war" again. Why was that a good choice?*
6. "Ms. Smythe *(nervously yelling)*—Stoney!" *Notice that when Ms. Smythe became nervous and yelled, other students became nervous as well. What else might she have done here?*
7. "I really think I need my inhaler." *Ms. Smythe recognized Syrie may have been in danger and, deciding she was not in imminent danger, left her place to get Syrie's inhaler.*
8. "What if Stoney comes in . . . and attacks us!" This scene provides a great opportunity to talk about the importance of not jumping to conclusions and avoiding hysteria; the importance of remaining calm.

4. *Poststory discussion:*
   Spend time discussing with students all the decisions that were quickly made, both by students and Ms. Smythe. You may wish to have students create a list of these decisions and discuss which they agree with and what alternative choices they might have made.

Lesson Conclusion:

Let students talk about how they think they might react to this event. Share ideas.

Lesson Extension:

The classroom teacher or librarian may wish to assign students to write a similar emergency preparedness skit in small groups.

Student Informal Assessment/Student Reflection:

The purpose of this skit is for students to experience an emergency situation that turned out not to be dangerous. Following this reading and discussion, the librarian may ask students to share—either orally or written—their overall reaction.

Professional Reflection:

Indicators of this lesson's success include:

- Students' level of engagement in discussion and reading.
- Students share additional questions or information about potential emergencies.

DISASTER PREPAREDNESS: THE UNEXPLAINED EXPLOSION

Characters:

Ms. Smythe: the school librarian
Principal James: the school principal

Fourth Grade Students:

Sam: the class safety officer
Syrie: a student with highly reactive allergies
Stoney: a quiet, mysterious, somewhat shy student
Pierce: always has the lead in the school play and tends toward the dramatic
Stella: science fair winner from last year
Petra: moved to the United States recently from her Eastern European wartorn country of birth
Hernandez: working on the science fair project as Sven's partner
Sven: working on the science fair project as Hernandez's partner
Jonathan: a history buff
Susan: a student with excellent computer skills

Scene:

Narrator: Groups of fourth grade students are in the library doing research for their science fair projects. Ms. Smythe, the only adult in the library, is working with one small group of students helping them locate information at computers. The other students are working at desks, writing, working at independent computer stations, and reading books and other material for their science projects in the reading corner.

Suddenly an enormous explosive sound from outside is heard. Immediately the lights flash, the computers black out, and all electricity is gone. Petra is frozen with fear. Pierce screams. All students' eyes turn toward Ms. Smythe.

Ms. Smythe: *[speaking quickly]* Boys and girls, I'm not sure what that explosion was but I think it's critical that, in case it was an earthquake or some event that is not over, we need to take cover. Don't rush; try to find a desk. Get under the desk and cover your heads.

Sam: *[pointing to a tall bookshelf]* But don't try to find cover near those social studies nonfiction shelves. Stay away from bookshelves that are higher than your head. *(Turning to Ms. Smythe.)* I am the head of the safety committee for our classroom, Ms. Smythe, I can help.

Ms. Smythe: *[as she rushes to help students duck under desks]* Good to know, thank you, Sam, please take cover.

Sam: *[turning to tell all his classmates]* If you can't find a desk to get under, you can stand in the doorway for protection.

Stella: Oh no! My science project! I didn't have time to save on the computer!

Sven: *[ducking under his desk]* We could have much bigger problems than not saving your work!

No one notices that Stoney is not in the room.

Susan: These computers are new enough they'll save your document automatically, with auto recovery. I'll help you. *(Susan helps Stella get under desks.)*

Ms. Smythe: *[getting the last student tucked under a desk]* Stay under the desks until we know something.

Petra: *[very nervously from under a desk]* Could we be attacked? It sounded like an attack! Could we be in war?

Jonathan: We have never been attacked on our own country before . . .

Petra: *[simultaneously]* But we could . . .

Hernandez: *[simultaneously]* What about 9/11? That was an attack.

Ms. Smythe: *[simultaneously]* Stay down.

Jonathan: *[from under his desk]* That was by a group of extremists, not another country attacking us.

Ms. Smythe: Let's not speculate until we hear something. Quiet everyone, please.

All students are now under desks and from her desktop Ms. Smythe has taken the list of students who are in attendance in the library.

Ms. Smythe: I just want to check on each of you. I have a list of who's supposed to be here. Don't get out from under your desk but just answer a question or two as I call your name. Sam, I may call on you to help.

Sam: Okay. We've had drills. We're pretty prepared, I think.

Petra: But that was lockdown drills and fire drills. We haven't had a war drill to prepare for being under attack. What do we do?

Ms. Smythe: Petra, it's okay. That loud noise did not sound like an attack or invasion. I'm pretty sure of that.

Petra: But we should have drills for war. We had drills in my country . . .

Ms. Smythe: Let me check on everyone first, Petra, and then we'll talk about it. *(To the room.)* Jonathan? Where are you? Are you under a desk? How are you feeling?

Jonathan: *[from under his desk]* Yes. I'm okay.

Ms. Smythe: Syrie, where are you?

Syrie: Here, Ms. Smythe, in the back.

Ms. Smythe: Are you under a desk? Are you alone? And are you okay?

Syrie: Yes.

Ms. Smythe: Yes, to . . ..

Syrie: Yes, I am under a desk, alone. I am okay but I should get my inhaler just in case. I have allergies. They get worse if I'm nervous.

Ms. Smythe: Where is your inhaler?

Syrie: It's in my backpack by the door.

Ms. Smythe: *[going toward the door then changing her mind]* Let me finish checking where everyone is. Then I'll get your backpack, Syrie. *(Turning attention to the rest.)* Petra, are you under a desk?

(No answer at first.)

Ms. Smythe: *[louder]* Petra, where are you?

Petra: Here. I'm here. I'm . . . What was that? What is happening?

Ms. Smythe: Petra, are you under a desk? Where? I will come there.

Petra: By the nonfiction.

(Ms. Smythe quickly moves to join Petra.)

Ms. Smythe: *[reaching Petra]* Petra, don't worry. We are all okay. There's no other noise. Nothing else we have heard. It was probably something that just happened once and now it's over. Are you okay? Would you like someone else to stay here by you? Sam, would you come over here and stay near Petra? *(Returning to under her desk.)*

Sam: *[leaving his desk and joining Petra]* Petra, we read about this in our safety meetings. There are way more emergencies that are just some kind of accidents than anything you mentioned.

Ms. Smythe: Thank you so much, Sam. *(Turning back to the rest.)* Stella, I can see you next to Susan. Susan and Stella, are the two of you feeling all right?

Stella and Susan: *[simultaneously]* Yes. We're here. We're fine.

Ms. Smythe: Great, you two, stay together, will you? *(Turning to the rest of the room and speaking loudly from under her desk.)* Sven, are you alone? Are you okay?

Sven: Yes, I'm here next to Hernandez. The two of us are discussing our science fair project.

Hernandez: Yes, we had a really good idea, but now I think are going to change it to something that has to do with preparing for emergencies.

Sven: Yeah, we were thinking we would make simulations of a volcano, an earthquake, and maybe a hurricane and compare the ...

Ms. Smythe: Wonderful! Great distraction. *(To the room.)* Stoney, are you here?

(No answer.)

Ms. Smythe: Stoney?

(Still no answer.)

Ms. Smythe: *[again, louder]* Stoney? ... Has anyone seen Stoney?

Sam: *[from his place by Petra]* When there's someone missing, we're supposed to report it right away to the office, Ms. Smythe.

Ms. Smythe: *[nervously yelling]* Stoney! Where are you? *(to herself)* Maybe he wasn't here when the explosion happened. *(to the group)* Did any of you see Stoney come to the library?

Syrie: Ms. Smythe, I really think I need my inhaler. *(somewhat breathless)* It's getting harder to breathe.

Ms. Smythe: *[heading toward the pile of backpacks near the door]* Okay, Syrie, don't panic. We'll get your inhaler. What does your backpack look like?

Susan: I really need to call my mom. She told me if there's ever an emergency at school I need to call her right away. She'll come and get me. She must be wondering what happened and why I haven't called.

Ms. Smythe: No, I ...

Petra: *[interrupting]* Maybe Stoney set off a bomb in the school. Maybe that's why he's not here!

Ms. Smythe: Oh no, Petra, I ...

Jonathan: *[interrupting]* You know ... in class Stoney was talking about a really violent movie he saw last night. And he said the bombings were really cool.

Pierce: Oh yeah! That's right! Stoney was talking about the violence in the movie, and he sounded like he was really into it!

Ms. Smythe:	Jonathan, Pierce, stop. Everyone, we are not going to jump to any conclusions, and we certainly will not assume anything about your classmate who is not here to defend himself. I'm going to go report Stoney missing and see if I can find out anything else. (*She gets out from under her desk and goes toward the phone near the checkout computer.)*
Sam:	You shouldn't come out until someone gives an all clear sign.
Jonathan:	Yeah! What if Stoney comes in and ...
Pierce:	...and attacks us!
Susan:	Ms. Smythe, if you grab Syrie's backpack to get her inhaler mine is right next to hers and I have a cell phone. You can take my cellphone under your desk and call someone.
Ms. Smythe:	Susan, you're a lifesaver. Thank you for your levelheadedness and good advice.
Sam:	She's right, the most important thing is staying calm.
Susan:	Mine is the black backpack with green straps.
Syrie:	And mine is the blue backpack right next to hers. But I'm better; I might not need my inhaler.
Ms. Smythe:	*[reaches for the backpacks and pulls them under the desk where she's covering]* Here's the phone. Here's your inhaler, Syrie.
Narrator:	All of this response happened in a matter of only 10 minutes. Just as Ms. Smythe is about to call the main office, Stoney walks back into the library. In his hand is the library bathroom pass. Stoney looks around curiously at everyone under the desks.
Ms. Smythe:	Stoney! You're okay! Where have you been?
Stoney:	*[a bit embarrassed]* Bathroom pass.
Ms. Smythe:	Did you notice other students in the hall?
Stoney:	No, I only went from here to the bathroom right next door.
Ms. Smythe:	But there were no other students in the hall?
Stoney:	No. What's going on?
Ms. Smythe:	*[suddenly realizing Stoney is unprotected]* Would you get under a desk while I call the office. (*She begins dialing just as the door opens and Principal James enters.)*
Principal James:	*[noticing all students under desks]* Excellent! Great job, fourth graders and Ms. Smythe. Is everyone accounted for?
Ms. Smythe:	Now, yes. What's going on?
Principal James:	You can all come out now from under the desk. *(As they come out, she continues explaining.)* Everything is under control. There was a major explosion at the Samson Factory nearby, but fire trucks are on the scene, and the minimal fires caused from the explosion have been extinguished. I see you have one of our safety officers, Sam, here. Thanks, everyone, you too, Sam. Return to what you were doing before; I have a few more rooms to check.
Ms. Smythe:	Thank you. We are all relieved. Yes, Sam was a great help. I think you should also recruit Susan to be on the safety committee.

The End

EMERGENCY PREPAREDNESS SCAVENGER HUNT

Lesson: "Finding" Emergency Prep

Grade Levels: 3–8, adaptable beyond

Time: 40 minutes

Introduction:

Often students simply follow rules for emergency drills but never discuss the "why's" of what is enacted. The purpose of this lesson is greater involvement of students in emergency preparedness. This lesson introduces key emergency preparedness supplies and/or actions. This lesson may be taught in response to a school having experienced a disaster or as part of the emergency preparedness curriculum.

Lesson Objectives:

- Identifying essential emergency preparedness supplies and actions
- Learning use of emergency preparedness supplies and purpose of actions

NET Standards:

2.d: Contribute to project teams to produce original works or solve problems (Student standard)

Resources, All:

Emergency preparedness websites (see "DOT GOV" Emergency Agencies worksheet for a list of federal emergency agencies)

Resources, Alternative #1:

Lists of emergency prep scavenger hunt items specific to the school or standards such as water, snacks, batteries, hand sanitizer, latex gloves

Resources, Alternative #2:

Lists of scavenger hunt emergency activities students must perform such as:

- Location of safety equipment (e.g., fire extinguisher)
- Contact incident commander to inform of emergency
- Procure contact information for local first responders
- Note protocol for evacuation

Librarian Instructional Role:

This lesson may be taught as a single lesson or as a part of the emergency preparedness curriculum.

Lesson Plan Enactment:

Lesson Introduction:

Discussion: Begin by asking students to discuss the following. (Record their answers for viewing and discussion. Add their posthunt results.)

When we perform drills—lockdown, fire drills, earthquake drills—there are procedures and resources necessary for successful drills.

What are some of the procedures we follow?

What resources do we use during drills?

How do resources differ between lockdown and fire drills? Earthquake drills?

(This discussion serves both as a good review of students' knowledge of emergency procedures and their understanding of what to do during drills.)

Lesson Procedure:

Begin by explaining the following:

Today we are going to review these emergency procedures and, more important, discuss why *each is important. We are going to play a game, Emergency Scavenger Hunt, in which teams will be responsible for finding resources to use in an emergency.*

Review basic scavenger hunt procedures if necessary.

After you have gathered your items, come back to your tables, write down your items, and then write a description of what purpose each item serves for emergency preparedness. Use emergency websites (see "DOT GOV Emergency Agencies worksheet" for a list of federal agencies) *for further details.*

You may wish to designate one type of emergency for which the school community needs practice and hone items and activities accordingly.

Quickly group students in teams of two to four. Review the norms for scavenger hunts and apply library rules as appropriate. Give students a list of the scavenger hunt items. Allow students time to gather emergency resources.

Scavenger Hunt, Alternative 1:

Have students search for the items necessary in an emergency. Items will vary by school and/or emergency, but the basic purpose is for students to understand how each item is used in an emergency and why each is important for safety.

Scavenger Hunt, Alternative 2:

Instead of giving students a list of items, let them first brainstorm items they believe would be valuable in a real emergency situation and in the posthunt writing have them record what each item would contribute in an emergency.

Scavenger Hunt, Alternative 3:

Instead of actual items, have students perform emergency prep activities necessary in an emergency. Again, activities will vary; the basic purpose is for students to understand what the activities contribute to student safety.

Lesson Conclusion:

Students share results and discuss the importance of having proper preparation for emergencies. Students may also create a poster describing one important emergency supply or action.

Student Assessment/Student Reflection:

Students' responses to the purpose of items found reveal their understanding of parameters of emergency preparedness.

Professional Reflection:

An indicator of this lesson's success is student understanding of emergency procedures and a call to action for accepting their role in school safety.

RECOVERY AND RENEWAL: "THE RED LEAF"

Lesson: "The Red Leaf," from *Lost & Found* by Shaun Tan

Grade Levels: 4-6 (adaptable for middle school)

Time: Varies depending upon extensions, 40 minutes minimum

Introduction:

This story provides multiple opportunities for discussion around **themes** such as:

- Loss and recovery
- Feelings of hopelessness or depression followed by hopefulness
- Understanding emotions and the power of emotional response

This lesson plan is open ended for various interpretations, just like the story. Librarians may read the story following a disaster or crisis in the class or school, or the story may be used to study **literary theme**. Tan provides an excellent perspective on one character's deep sense of sadness, and he has left the emotional reactions open to interpretation.

Lesson Objectives:

- Giving students an opportunity to self-reflect on a crisis in their lives
- Providing a springboard for class discussions during disturbing times
- Understanding how a story's themes are open to interpretation

Common Core Standards:

CCSS.ELA-LITERACY.RL.5.2: Determine a theme of a story, drama, or poem from details in the text, including how characters in a story or drama respond to challenges or how the speaker in a poem reflects upon a topic; summarize the text.

Resources:

Tan, Shaun. *Lost & Found.* New York: A. Levine, 2011.

Outline/trace of full-page red maple leaf

YouTube video of "The Red Tree," music by Prema Yin:

https://www.youtube.com/watch?v=wAsEq6EG8jE

Instructional Roles:

This lesson can be taught in isolation as a single library lesson or in collaboration with classroom teachers as a literature lesson (theme) or a problem-solving lesson.

Use this story also to teach other literary forms and perspectives:

- Compare and contrast
- Metaphors
- Symbolism
- Tone

Lesson Plan Enactment:

Prestory discussion:

Begin by posing the following questions: *A person can feel "lost" in a lot of different ways. Can you describe a time when you felt lost? Sometimes we can identify exactly what is making us feel this way. At other times, we may feel sad, lost, alienated, or alone and not necessarily understand why.*

Lesson Procedure and Assignment:

1. Read "The Red Tree," the first story in Tan's book Lost and Found.
2. The following questions may prompt a deeper discussion:
    - *What details in the illustrations illuminate the girl's sadness and hopelessness?*
    - *What are ways the author uses the landscapes in the story to show emotion?*
    - If students in the school have suffered a loss, discuss the author's perspective: *In the author's note following this story, Tan explains that he "wanted to create something useful from what can seem to be a useless experience—an abject feeling of hopelessness—but more important, to simply acknowledge its reality." Tan also states, "Most consistently positive comments [about this story] have come from children."*
    - *What do you think the tree symbolize?*
3. View one or more versions of the story that have been set to music and have been animated. Ask, *How does the background music in the video complement the tone of the story?*
4. Word wall assignment: Hand out a blank red leaf. Remind students that in the short story and the YouTube video of "The Red Tree," a young girl has an overwhelming sense of loss and alienation. Ask students to find details in the story that demonstrate the author's way of expressing emotions. Ask them to tell which descriptive words express these feelings—words beyond the text? Write them on the red leaf.
5. Theme discussion: In his author's note, Tan writes about one of the themes in the book, "important things in life are not always immediately visible." Let students discuss in small groups what they think the **theme** of this story is. Share and compare.
6. Theme assignment: Ask students to use their list of "feeling words" to describe the emotions portrayed in the story, their group's discussion ideas, and the story itself to determine the theme of the story. Ask them to write one or two sentences to describe what they think is the theme of the story.
7. Librarians and teachers may wish to have students share or post themes.

Lesson Conclusion:

Whether this lesson is used in response to a critical incident in the school or simply for studying theme, the story's end warrants a final discussion on the optimistic conclusion. Librarians may wish also to conclude the lesson by reading Tan's author's note.

Lesson Extension: Recovery

1. Have students turn over their red leaves and write "recovery words" or words that describe the red tree.
2. Post all the red leaves onto a "feeling-healing tree" on a bulletin board or in the hallway.
3. Have the students compose a poem describing how they interpret this story.
4. Ask students to illustrate the poems. Then post them in the library.

Student Assessment/Student Reflection:

Student "Theme" assignment pose reflections on the theme of "The Red Tree."

Professional Reflection:

Indicators of this lesson's success include:

- Students' "red leaf" word wall
- Students checking out this and other books on similar subjects
- Students' presentations reflecting an interest in a call-to-action

Additional Recommended Resources and Lesson Extensions:

Although the titles listed below are for younger readers, librarians may wish to show how each complement both the plot and the tone of Tan's stories:

- *The Little Yellow Leaf* by Carin Berger
- *Beegu* by Alexis Deacon
- *The Scar* by Charlotte Moundlic

The YouTube video of "The Lost Thing" can be found here:

https://www.youtube.com/watch?v=t1YG7ZXfC6g

This Academy Award winning video animates Tan's story.

Teacher-librarians may wish to view this YouTube video and read "Lost and Found," as the two stories have several interesting parallels, including the use of the color red in illustrations to evoke emotion.

An excellent YouTube video in which Shaun Tan is interviewed about his newest book, *Rules of Summer*, can be found here:

http://www.youtube.com/watch?v=fbjMpWc3amk

SEPTEMBER 11, 2001

Lesson: Hope, Resilience, and Renewal

Grade Levels: 4–7

Time: 2–4 days, 30 minutes each day

Introduction:

This lesson introduces students to the events of 9/11 through stories of hope, resilience, and renewal.

Lesson Objectives:

- Identifying heroes and heroic acts from September 11, 2001
- Understanding the resilience of those impacted by 9/11

Common Core Standards:

ELA-RI.6.2: Determine a central idea of a text and how it is conveyed through particular details; provide a summary of the text distinct from personal opinions or judgments.

ELA-W.6.3.A: Engage and orient the reader by establishing a context and introducing a narrator and/or characters; organize an event sequence that unfolds naturally and logically.

NET Standards:

3.B: Locate, organize, analyze, evaluate, synthesize, and ethically use information from a variety of sources and media.

Resources: Internet access

http://www.911memorial.org

http://voicesofsept11.org

http://www.rmsc.org/Data/Documents/newsandprograms/NP_fall_2011_final.pdf

Librarian's Instructional Role:

This lesson may be taught in response to a school having experienced a disaster or as part of a history or current events unit, Memorial Day curriculum, or a crisis or event anniversary.

Lesson Plan Enactment:

Lesson Introduction:

Discussion: Begin by asking what students currently know about the events of 9/11. Read one of the stories listed in the "Additional Resources" section of this lesson plan to spark discussion of responses to 9/11.

Lesson Directions:

In light of tragic events, it is important for students to understand how everyday heroes rise to such occasions and communities are drawn together. In this lesson students will work with a partner to find information about the rescues, other heroic acts, recovery, and resilience related to 9/11 events.

Lesson Procedure:

1. Pair students for research.
2. On document camera, show students links to 9/11 research.

3. In pairs, have students browse various 9/11 sites to find a story of heroism or resilience that is of interest to them.
4. Have students select one such story (during and in response to 9/11) and read the information they have found.
5. Students should then select key terms or words from the 9/11 heroic story they found and search online for additional information.
6. Students should find a minimum of three sources on their topic/hero.
7. Have students take notes on key events of the heroic act.
8. Let students select one of the following alternatives to present their hero.

Heroic Acts, Alternative 1:

Write a paragraph describing the hero or heroic act you read about.

Heroic Acts, Alternative 2:

Write a letter to the hero in praise of his or her heroism.

Heroic Acts, Alternative 3:

Write a poem in which the act of heroism is described through the eyes of a witness or recipient of the act.

Heroic Acts, Alternative 4:

Create a memorial to honor the hero. This may be designed as a statue, a work of art, or an award recounting the act of heroism.

Lesson Conclusion:

After all students have created their responses, display the works in the library.

Student Assessment/Student Reflection:

Students' assignments demonstrate an understanding of the importance of viewing such events through acts of resilience and heroism rather than just the attack.

Professional Reflection:

Students' work that shows insight into the importance of honoring everyday heroes is an indicator of this lesson's success.

SEPTEMBER 11, 2001 STORIES

Included are nonfiction and some fiction titles based on true stories. The primary selection criteria was to provide balanced but resilience-focused titles that provide positive perspectives on 9/11.

Benoit, Peter. *September 11 Then and Now.* Scholastic, 2012.

The initial chapters report 9/11 events with Al Qaeda/bin Laden background; the remaining chapters present response, political action, healing, and details of Navy SEALS killing of bin Laden. Extensive bibliography, websites, glossary, and index.

Brown, Don. *America Is under Attack: The Day the Towers Fell.* Roaring Brook, 2011.

This text fills the need for a book for elementary students without the severe images. Text presents facts and overview; muted pastel drawings minimize 9/11 horror.

Curtiss, A. B. *The Little Chapel That Stood.* Old Castle, 2003.

St. Paul's Chapel stands across the street from where the Twin Towers once stood. The chapel remains, and this tribute tells the story of the World Trade Center attack in poetry and gentle pastel illustrations.

Deedy, Carmen Agra. *14 Cows to America.* Peachtree, 2009.

When Kimeli returns from New York City to Kenya, he tells his people, once fierce warriors, the story of September 11. Elders invite members of the American embassy to their village, where they perform a sacred ceremony and present the United States with 14 cows.

Gerstein, Mordecai. *The Man Who Walked between the Towers.* Roaring Brook, 2003.

As construction of the World Trade Center nears completion, a French aerialist is determined to walk a tightrope between the towers. After he completes his mission he is arrested, but the judge sentences him to perform in the park for kids.

Jonell, Lynne. *Bravemole.* Putnam, 2002.

This fictionalized animal tale recounts the events of September 11 through the eyes of a family of moles whose lives are abruptly upended by a dragon attack. Mole works to assure Babymole that adult moles' heroic acts will keep them safe.

Kalman, Mairi. *Fireboat: The Heroic Adventures of the* John J. Harvey. Puffin, 2005.

Built in 1931, the fireboat *John J. Harvey* is initially the largest and fastest in New York Harbor. Following its retirement it was restored but was not used as a fireboat until 9/11, when it was called upon to fight fires for four straight days and nights.

Stuart, Gail D. *America under Attack: September 11, 2001.* Lucent, 2002.

This well-researched and documented account of events leading up to, during, and following the 9/11 attack includes many original photos, personal stories, primary sources, and an extensive bibliography. Insets include descriptions of problems of racial profiling at airport screening alongside many personal anecdotes.

Wheeler, Jill C. *September 11, 2001: The Day That Changed America.* ABDO, 2002.

The first 14 pages in this War on Terrorism series title present the terror of 9/11 in large, double-page, graphic color photos and few words. The remaining text presents details of the day and follow-up descriptions of heroism, leadership, and poignant anecdotes. Balanced reporting, glossary time line, and index.

Winter, Jeanette. *September Roses.* Farrar Strauss. 2004.

Based on a true story, this book provides global awareness of the world's response to 9/11. Two sisters from Africa arrive in New York City on September 11, bringing 2,400 roses for a flower show. Stranded, the sisters take the flowers to Union Square, where they design a tribute to the World Trade Center.

FEDERAL AGENCIES WORKING FOR YOU

Name______________________________ Date______________

Each of the federal agencies listed on this page has a role in providing U.S. citizens with protection. For each of the agencies listed, type the name of the agency in your computer's search bar, find the agency on the Internet, then record its home page address in the "Web Address" column and the abbreviation used in the "Acronym" column.

Example: United States Department of Education

Web Address: http://www.ed.gov Acronym: DOE

Agency	Web Address	Acronym
Bureau of Alcohol, Tobacco, Firearms and Explosives		
Centers for Disease Control and Prevention		
Federal Bureau of Investigation		
Federal Emergency Management System		
National Institutes of Health		
National Oceanic and Atmospheric Administration		
National Security Agency		
U.S. Department of Homeland Security		
Readiness and Emergency Management for Schools		
U.S. Secret Service		

1. Which agencies' websites were easiest to access? Why?
2. Were any websites difficult to find? What strategies did you use to find each website?
3. What do all these agencies have in common?

DOT GOV

Lesson: Using government websites as instructional resources

Grade Levels: 5–7 (adaptable for older audiences)

Time: Varies, 30 minutes each day over two to three days

Introduction:

Introducing emergency preparedness and disaster recovery website(s) in a lesson on URL .gov endings can provide students with tools for safety and recovery as well as an understanding of how to apply different types of government website information.

Lesson Objectives:

- Locating government agencies' websites
- Discovering strategies for determining if a website has relevant information
- Recognizing that multiple government agencies work to provide protection

Common Core Standards:

CCSS.ELA-LITERACY.RI.6.2: Determine a central idea of a text and how it is conveyed through particular details; provide a summary of the text distinct from personal opinions or judgments. (sixth grade)

Resources:

Computers

List of Internet sites to search

Websites Used:

Bureau of Alcohol, Tobacco, Firearms and Explosives

Centers for Disease Control and Prevention

Federal Bureau of Investigation

Federal Emergency Management System

National Institutes of Health

National Oceanic and Atmospheric Administration

National Security Agency

U.S. Department of Homeland Security

Readiness and Emergency Management for Schools (USDE)

U.S. Secret Service

Librarian's Instructional Roles:

This lesson has **two worksheets** and can be taught as two single information skills lessons or as part of a longer unit on emergency preparedness or a lesson using web resources.

Day one: Federal Agencies Working for You

Day two: Using Government Websites for Emergency Preparedness and Recovery

Lesson Plan Enactment:

Lesson Introduction:

- Distribute the "Federal Agencies Working for You" worksheet.
- It contains a list of government agencies. Have students see how quickly they can find the website for each agency on the list. Have students write the web addresses and acronyms for each government agency.
- Assign or discuss the three remaining questions at the bottom of the worksheet.

End of Worksheet Discussion: Read to class:

- *How did you find the websites?*
- *What strategies did you use to locate the web address?*

Lesson Procedure and Assignment:

Introduction: *Today we are going to look for ways to determine which government agency websites are appropriate for finding information you might need on emergency preparedness and recovery. Some of the same strategies you just used to find websites can be used to determine which government websites provide information you need.*

Here are several emergency preparedness scenarios for which you may need a government agency:

- How to make a home emergency preparedness kit.
- Great Grandma has no computer skills, and she needs your help in filling out forms to report the water damage that her house suffered in her town's recent flood.
- What is the government doing to protect you from terrorist attacks?
- How can I be a responsible, informed citizen about terrorism?
- As the school's student volunteer on the safety committee you want information on preparedness for potential school violence.
- What are the best hurricane or tornado preparations you can do?
- Where do you go to find out about weather patterns for potential storms in your region?
- If our school experiences an influenza epidemic, what are some strategies you can use to ensure you will stay healthy?

DIRECTIONS:

1. Distribute the "Using Government Websites for Emergency Preparedness and Recovery" worksheet.
2. *Read through this list of information needs ("Research Subject").*
3. *Determine what key word search you may need to use.*
4. *Write those suggested key words in the first column for each research subject.*
5. *Select one to two websites ("Gov. Site Selected") you think will have the information you are looking for.*
6. *Go to those websites and type your selected key words in the search bar.*
7. *The website may send you to another page. If so, write the new page in the third column ("List that New Page").*
8. If you think you have found the page you need, skim the initial text and see if you are able to get the information needed. If so, add *"yes"* for this page to column 3.

Follow-Up Discussion Questions:

How do you know you have found an accurate website?

Why is this information likely more trustworthy than a .com website?

Discussing Differences:

Differences between government agencies or departments and commercial ones:

- *How can you tell if a website is sponsored by a government agency?*
- *What information do you need to first find the authentic website of a particular government agency?*
- *When might you want to access a government agency?*
- *What advantage might a government site have over a commercial one?*

Lesson Conclusion:

Concluding remarks: *Notice that multiple government agencies work to keep U.S. citizens safe and to assist them when disasters strike. These agencies also provide valuable information citizens can read on their own to prepare for or recover from disasters.*

Student Assessment/Student Reflection:

Librarians can determine if students learned lesson objectives:

Assessment of student worksheets:

Federal Agencies Working for You worksheet

Using Government Websites for Emergency Preparedness worksheet

Professional Reflection:

Indicators of this lesson's success include students' active participation (qualitative) and assignment results (quantitative).

USING GOVERNMENT WEBSITES FOR EMERGENCY PREPAREDNESS AND RECOVERY

Name______________________________Date__________________

Read through this list of emergency preparedness situations. Then look up the government agencies on the list and decide which agency will provide the information you need to answer your research needs.

Research Subject	Keyword	Govt. Site(s) selected	Sent to a New Page? List that New Page
What supplies do you need to make a home emergency preparedness kit?	"Emergency prep kit"	FEMA	http://www.ready.gov/kit YES
Great grandma has no computer skills and she needs your help in filling out forms to report the flood damage her house suffered in her town's recent flood.			
What is the government doing to protect you from terrorist attacks?			
What can I do to be a responsible and informed citizen about terrorism?			
As the school's student safety committee rep you need information on preparedness for potential school violence.			
What should I know to prepare for hurricanes or tornadoes?			
Where do you look to find out about weather patterns for potential storms in my region?			
If our school experiences an influenza epidemic, what are some strategies you can use to ensure you will stay healthy?			

Bureau of Alcohol, Tobacco, Firearms and Explosives
Center for Disease Control
Federal Bureau of Investigation
Federal Emergency Management System
National Institute of Health
National Oceanic and Atmospheric Administration
National Security Agency
United States Department of Homeland Security
Readiness and Emergency Management for Schools (USDE)
United States Secret Service

THE HEART OF A HURRICANE

Lesson: Writing Hurricane Poetry

Grade Levels: Middle school and high school (adaptable for younger audiences)

Time: 45 minutes or one class period, more if extensions are used

Introduction:

Students who have experienced a hurricane often need an outlet, that is, a way to express their post hurricane feelings. Writing poetry can provide such an outlet for some students.

Lesson Objectives:

- Understanding that poetry on one subject can be expressed through multiple forms of poetry.
- Writing poetry as a method of expressing feelings, providing an emotional outlet
- Reading, interpreting, and writing various forms of poetry on one subject

Common Core Standards:

CCSS.ELA-LITERACY.RL.7.5: Analyze how a drama's or poem's form or structure (e.g., soliloquy, sonnet) contributes to its meaning

CCSS.ELA-LITERACY.W.7.3.D: Use precise words and phrases, relevant descriptive details, and sensory language to capture the action and convey experiences and events.

CCSS.ELA-LITERACY.W.9-10.3.D: Use precise words and phrases, telling details, and sensory language to convey a vivid picture of the experiences, events, setting, and/or characters.

Resources:

Access hurricane poems online, print copies for student use (below).

"The Hurricane" by William Cullen Bryant (1854)

"Hurricanes" by Nicholas Peters

http://www.poemhunter.com/poem/hurricanes-4/

Poetry Soup provides excellent examples of hurricane poems, and some of them are responses to very recent hurricanes (e.g.., "Megan in Memory of Hurricane Sandy").

Note: not all these Poetry Soup poems are appropriate for all audiences.

http://www.poetrysoup.com/poems/best/hurricane

http://hellopoetry.com/words/16535/hurricanes/poems/

Librarian's Instructional Role:

For this lesson the librarian may take the lead and teach an individual lesson, or the lesson may be part of a longer poetry unit taught in collaboration with classroom teacher(s).

Lesson Plan Enactment:

Lesson Introduction: Everyone who has experienced a hurricane has his or her own personal emotional response to that experience. Today we are going to look at how different poems written on that subject, hurricanes, express emotional response to hurricanes in very different ways. We will read a few together, then I will give you a chance to select and read different poems on your own, and finally we will/may try our hand at writing a poem about hurricanes in a chosen form. As you read or listen to these poems, try to imagine what the poet is feeling. Can you relate to this poet's hurricane experience and response?

Lesson Procedure and Assignment:

1. Poetry reading and discussion:

You may wish to begin by reading the following classic poems about *hurricanes* to the whole class:

The Hurricane by
William Carlos Williams
The tree lay down
on the garage roof
and stretched, You
have your heaven,
it said, go to it.

From Shakespeare's *King Lear:*

Blow, winds, and crack your cheeks! Rage! Blow!

You cataracts and hurricanoes, spout

Till you have drench'd our steeples, drown'd the cocks!

You sulphurous and thought-executing fires,

Vaunt-couriers to oak-cleaving thunderbolts,

Singe my white head! And though, all-shaking thunder,

Smite flat the thick rotundity o' the world!

Crack nature's moulds, all germens spill at once,

That make ingrateful man!

You may wish to discuss/compare the way the two writers express their response. What emotions are evoked in *King Lear*?

2. *Prepare and read poems*: Distribute various hurricane poems (see the resources section in this lesson plan) for students to read. Give students the opportunity to read several poems about or in response to hurricanes. They may read silently or aloud to each other.

3. *Introduction to assignment*: After students have had some time to read and discuss responses to various poems on their own, let them discuss and share the various poets' emotional responses to hurricanes through their poetry. Let them share their thoughts and responses related to the hurricane poems.

Librarians may ask:

- *A great number of poets have written poems about hurricanes. Why do you think this is so?*
- *Were there poems you read where the poet seemed to have the same reaction to hurricanes as you? What was similar?*
- *What stood out in the hurricane poems you read?*
- *What similarities did you notice across more than one poem?*

4. Assignment:

After students have had a chance to read and discuss some of the poems, give them an opportunity to write their own hurricane poem. Some may wish to illustrate their poem.

5. Alternative 1:

You may wish to assign all students to write a hurricane poem using the same format. Review some of the more familiar poetry forms such as haiku, acrostic, or free verse.

6. Alternative 2:

Rather than assign all students to write using the same format, leave the form open to student choice. For example, some may wish to write with a partner (poems of 2 voices).

Lesson Conclusion:

Let students talk about their own hurricane response. Let students read their poems.

Lesson Extension:

Have students write a poem in the shape of a hurricane about their experience or their emotional response to a hurricane.

Student Assessment/Student Reflection:

Criteria established for student poems

Details of student response during conclusion: *Do you think this is a beneficial way to release emotions you have regarding hurricanes?*

Professional Reflection:

Indicators of this lesson's success include:

- Students' level of engagement in writing poetry
- Students' follow-up discussion, interest, and library check out (poetry)
- Students share additional questions or information about their process in writing poetry.

Additional Resources: Hurricane Poetry Books

Watson, Rene. *A Place where Hurricanes Happen*. New York: Random House Books for Young Readers, 2010.

Engle, Margarita. *Hurricane Dancers: The First Caribbean Pirate Shipwreck*. New York: Henry Holt, 2010.

Smith, Patricia. *Blood Dazzler*. Minneapolis: Coffee House, 2008. (use selectively, as it is not appropriate for all audiences)

Bibliography

AboutOurKids.org. Chapter 7: Addressing tough topics and questions: Talking to children about traumatic events. (2010). New York Child Study Center. Retrieved online September 1, 2014, at http://www.aboutourkids.org/files/articles/addressing_tough_topics_06.pdf

Abramsky, S. (2013, July 17). Dust bowl blues: A severe drought in the Southwest is devastating crops and farm communities—and sending a warning about climate change. *Nation*. Retrieved online January 22, 2014, at http://www.thenation.com/article/175318/dust-bowl-blues?page=0,1

Active Shooter Awareness Virtual Roundtable Webinar. (2011). Department of Homeland Security.

Alire, C. A. (2000). *Library Disaster Planning and Recovery Handbook*. Chicago, IL: Neal-Schumann.

Allen, J., Allen, S. F., Latrobe, K. H., Brand, M., Pfefferbaum, B., Elledge, B., Burton, T., & Guffey, M. (2012). The power of story: The role of bibliotherapy for the library. *Children and Literature,* Spring, 2012, pp. 44–49.

Amarasiri, U. (2005). Rising from the wreckage: Development of tsunami-affected libraries in Sri Lanka. *IFLA Journal, 31*(4), 307–314.

American Academy of Pediatrics. *Guidelines for responding to the death of a student or school staff.* National Center for School Crisis and Bereavement. Retrieved online January 14, 2014, at http://www.stchristophershospital.com/userfiles/Guidelines%20for%20Responding%20to%20the%20Death%20of%20a%20Student%20or%20School%20Staff%20Jan%202013.pdf

Anderson, C. A., Shibuya, A., Ihori, N., Swing, E. L., Bushman, B. J., Sakamoto, A., Rothstein, H. R., Saleem, M., & Barlett, C. P. (2010). Violent video game effects on aggression, empathy, and pro-social behavior in Eastern and Western countries: A meta-analytic review. *Psychological Bulletin, 136*(2), 151–173.

Anderson, G. A. (2008). Emergency use of cell phones in school: Yes. *EducationSpace 360.* Retrieved online January 7, 2014, at http://www.educationspace360.com/index.php/emergency-use-of-cell-phones-in-school-yes-22636/

Arias, E. (2009). *National vital statistics report: U.S. life tables, 2009.* Centers for Disease Control and Prevention. Retrieved online March 31, 2014, at http://www.cdc.gov/nchs/nvss.htm

Baggerly, J., & Exum, H. A. (2008). Counseling children after natural disasters: Guidance for family therapists. *American Journal of Family Therapy, 36,* 79–93.

Barack, L. (2011, September 8). Media specialists help Joplin, MO, rebuild school libraries following deadly tornado. *School Library Journal* [Internet].

Barron, D. (1997). Violence, schools, society, and the school library media specialist. *School Library Media Activities Monthly,* (13), 47–50.

Barron, D. (1999). Violence and the school library media specialist. *School Library Media Activities Monthly, 1*(16) 48+.

Bartley, N., & Armstrong, K. (2014, March 24). Site has long history of slide problems. *Seattle Times.*

Bedford, A. W. (2006). Children's literature in a post-Katrina world. *Language Arts, 84*(3), 300.

Berman, M. (2014, March 24). Everything you need to know about the Washington landslide. *Washington Post.*

Biag, Manuelito. (2014). Perceived school safety: Visual narratives from the middle grades. *Journal of School Violence.* 13(2), 165–187.

Blad, E. (2013a). Report of the state's attorney for the judicial district of Danbury on the shooting at Sandy Hook Elementary School and 36 Yogananda Street. *Education Week, 33*(13), 4.

Blad, E. (2013b). Sandy Hook: Words and actions. *Education Week, 33*(14), 18–19.

Boyse, K. (2010). *Television and children.* University of Michigan Health Services. Retrieved online May 1, 2014, at https://www.med.umich.edu/yourchild/topics/tv.htm

Breslin, S. (2104, March 18). Tornado-stricken Oklahoma town invests in helmets for storm protection in schools. *Weather Channel.* Retrieved online March 24, 2014, at http://www.weather.com/safety/tornadoes/moore-schools-helmets-donated-20140317

Brock, S., Lazurus, P., & Jimerson, S. (2002). *Best practices in school crisis prevention and intervention.* Bethesda, MD: National Association of School Psychologists.

Bushman, B. J., & Huesmann, L. (2012). Effects of violent media on aggression. In D. G. Singer & J. L. Singer (Eds.), *Handbook of children and the media* (2nd ed., pp. 231–248). Thousand Oaks, CA: Sage.

Campbell, R. (2014). School Fires, 2007–2011. *Educational properties, National fire prevention association.* Retrieved online March 12, 2014 at https://www.nfpa.org/safety-information/for-consumers/occupancies/school-fires

Candiotti, S., & Botelho, G. (2014, January 18). *2 students shot in Philadelphia high school gym.* CNN. Retrieved online January 26, 2014, at http://www.cnn.com/2014/01/17/justice/philadelphia-high-school-shooting/

Church, A. (2013). Tapping into the skills of school librarians. *Principal Leadership, 14*(3), 44–46.

Class, P. (2009, June 8). At last: Facing down bullies (and their enablers). *New York Times.*

Coatney, S. (Ed.). (2010). *The many faces of school library leadership.* Santa Barbara, CA: Libraries Unlimited.

Coburn, A. W., Spence, R. J., & Pomonis, A. (1994). *Vulnerability and risk assessment.* Cambridge: Cambridge Architectural Research Limited. Retrieved online March 30, 2014, at http://worldbank.mrooms.net/file.php/356/2234/Introduction%20Reading%20-%20VulnerabilityAndRiskAssessmentGuide.pdf

Coe, J., Jeong, Y., & Cole. C. (2013, May 9). More explosive devices found at Tempe student's home. *Arizona Republic.*

Comfort, L. K. (2006). Communications and catastrophe: Building resilient communities for disaster response. In A. White & P. Eisinger (Eds.), *Cities at risk: Catastrophe, recovery and renewal in New York and New Orleans.* Milano: New School for Management and Urban Policy. Retrieved online February 11, 2014, at http://www.newschool.edu/milano/nycaffairs/documents/CitiesatRisk_Report.pdf

Connolly. (2013). Inside out: Reflections from Newtown. Retrieved online at https://reclaimingjournal.com/sites/default/files/journal-article-pdfs/22_3_Connelly.pdf

Costello, T., and Schecter, A. (2014, May 19). *Why aren't there tornado shelters in every Oklahoma school?* NBC News. Retrieved online May 20, 2014, at http://www.nbcnews.com/storyline/moore-tornado/why-arent-there-tornado-shelters-every-oklahoma-school-n107256

Council on School Health. (2008). Disaster planning for schools. *Journal of the American Academy of Pediatrics.* Retrieved online December 26, 2013, at http://pediatrics.aappublications.org/content/122/4/895.full.html

Cowan, K. C., & Paine, C. (2013). What really works. *Principal Leadership, 13*(7), 12–16.

Cowan, K. C., & Rossen, E. (2013). Responding to the unthinkable: School crisis response and recovery. *Phi Delta Kappan, 95*(4), 8–12.

Cowan, K. C., Vaillancourt, K., Rossen, E., & Pollitt, K. (2013). *A framework for safe and successful schools* (Brief). Bethesda, MD: National Association of School Psychologists.

Das, Sumi. (2014, May 15). *Text to 9-1-1 . . . for some.* CBS News. Retrieved online May 15, 2014, at http://www.ozarksfirst.com/story/d/story/text-to-9-1-1-for-some/20275/TAk9ueehLkSWbGCqbGjhDg

Dorn, C. (2013). *Safe Havens International blog.* Retrieved online December 9, 2014, at www.safehavensinternational.org/category/safe_havens_blog/

Doughton, S. (2013, October 15). Grays Harbor County School to build first U.S. vertical-tsunami refuge. *Seattle Times.*

Duststorm on the loose. (2014). *Kidsworld.* Retrieved online January 23, 2014 at http://www.kidzworld.com/article/707-dust-storm-on-the-loose

Ederhardt, G., Flagg, G., Goodes, P., & Kniffel, L. (2006). New Orleans gathering sends message of hope and renewal. *American Libraries, 37*(7), 42–63.

Emergency Alerts. (2014). Federal Emergency Management Administration. Retrieved online January 23, 2014, at http://www.ready.gov/alerts

Emergency Crisis Plan/Procedures. (2012–2013). Rock View Elementary School, Montgomery School District, Montgomery, Alabama. Retrieved online August 26, 2014, at http://www.montgomeryschoolsmd.org/uploadedFiles/schools/rockviewes/aboutus/WebParentEmerg(1).pdf

Emergency Response Crisis Management District Plan, 2009–2010, (2010). Stockton Unified School District. Retrieved online August 26, 2014 at http://rems.ed.gov/docs/repository/rems_000048_0011.pdf

Esper, J. (2014, April 21). Mitigating a school crisis: An all-hazards approach. *Partners in Emergency Preparedness Conference,* Tacoma, WA, April 15, 2014.

Everhart, N. (2013). Defining a vision of outstanding school libraries. *Teacher Librarian, 41*(1), 14–19.

Farrell, A., & Crimmins, D. (2001). Coping with disaster. *Exceptional Parent, 31*(11), 50–54.

Federman, J. (Ed.). (1998). *National Television Violence Study.* Center for Communication and Social Policy. Santa Barbara: University of California, Santa Barbara.

Fein, R. Pollack, W., Borum, R., Modzeeski, W., & Reddy, M. . (2002). *Threat assessment in schools: A guide to managing threatening situations and to creating safe school climates.* Washington, D.C.: U.S. Department of Education and U.S. Secret Service. Retrieved online January 21, 2014, at http://www.secretservice.gov/ntac/ssi_final_report.pdfs

Feinberg, T. (1999). The Midwest floods of 1993: Observations of a natural disaster. In A. S. Canter & S. A. Carroll (Eds.), *Crisis prevention and response: A collection of NASP resources* (pp. 223–239). Bethesda, MD: National Association of School Psychologists.

Ferguson, J., & Beal, T. (2013, November 10). Arizona's deadliest corridor for dust storms is 10 miles between Tucson and Phoenix. *Arizona Daily Star.*

Fortson, J. (1991). *Disaster planning and recovery: How-to-do-it manual for libraries.* Chicago: Neal-Schuman.

Frank, J. (2011). *The Impact of Hurricane Katrina on Gulf Coast libraries and their disaster planning.* Master's Thesis, San Jose State University: SJSU ScholarWorks.

Furlong, M., Felix, E. D., Sharkey, J. D., & Larson, J. (2005, September). Preventing school violence: A plan for safe and engaging schools. *Student Counseling,* pp. 11–15.

Gavigan, K. (2012). Caring through comics: Graphic novels and bibliotherapy for grades 6–12. *Knowledge Quest, 40* (5), 78–80.

Giannopoulou, I., Strouthos, M., Smith, P., Dikaiakou, A., Galanopoulou, V., & Yule, W. (2006). Post-traumatic stress reactions of children and adolescents exposed to the Athens 1999 earthquake. *European Psychiatry, 21,* 160–166.

Gilmore-See, J. (2010). *Simply indispensible: An action guide for school librarians.* Santa Barbara, CA: Libraries Unlimited.

Good, C., McIntosh, K., & Gietz, C. (2011, September/October). Integrating bullying prevention into schoolwide positive behavior support. *Council for Exceptional Children,* pp. 48–56.

Gordon, R., & Wraith, R. (1993). Responses of children and adolescents to disasters. In J. P. Wilson & B. Raphael (Eds.), *International handbook of traumatic stress syndromes.* New York: Philomel.

Griffiths, M. D., Davies, M. O., & Chappell, D. (2004). Online computer gaming: a comparison of adolescent and adult gamers. *Journal of Adolescence, 27*(1), 87–96.

Gross, T., & Clemens, S. G. (2002). Painting a tragedy: Young children processing the events of September 11, 2001. *Young Children, 57*(3), 44–51.

Guide to school vulnerability assessment report. (2008). U.S. Department of Education.

Haberty, S. (2014, January 24). 5 days of school shootings, lockdowns, scares jolt U.S. campuses. *Los Angeles Times.*

Halls, Kelly M. (2010, April). Biblio peace of mind. *VOYA: The Voice of Youth Advocacy,* pp. 34–35.

Halstead, D., & Jasper, R. (2005). *Disaster planning.* Chicago: Neal-Schuman.

Healy, J. (2014, January 16). In age of school shootings, lockdown is the new fire drill. *New York Times.*

Hebert, B. B., & Ballard, M. B. (2007). Children and trauma: A post-Katrina and Rita response. *Professional School Counseling, 11*(2), 140–144.

Heron, M. (2010). National vital statistics report: Leading causes of death. *Centers for Disease Control and Prevention, 62*(6). Retrieved online January 12, 2014, at http://www.cdc.gov/nchs/data/nvsr/nvsr62/nvsr62_06.pdf

Hibbert, Chuck. (2014, April 15). *Franklin Regional High School stabbing: Is pulling the fire alarm the best practice?* National School Safety and Security Services. Retrieved online May 1, 2014, at http://www.schoolsecurity.org/2014/04/franklin-regional-high-school-stabbing-pulling-fire-alarm-best-practice/

Hobbs, A. (2014, January 4). Two Olympia schools suffer damage from attempted arsons. *Olympian.* Retrieved online January 3, 2014, at http://www.theolympian.com/2014/01/04/2914985/two-olympia-schools-suffer-damage.html

Hodges, J., Bhattacharya, D., & Gray, J. (April, 2008). *Legal preparedness for school closures in response to pandemic influenza and other emergencies: A review and report submitted to the centers for disease control and prevention.* Center for Law and the Public's Health at Georgetown and John Hopkins Universities. Retrieved online March, 13, 2014, athttp://www.flu.gov/planning-preparedness/school/schoolclosures.pdf

Horne, J. (2011, Spring). New schools in New Orleans. *EducationNext, 11*(2). Retrieved online February 2, 2014, at http://educationnext.org/new-schools-in-new-orleans/

Hunsbuger, M. (1985). Teaching reading methods: How do preservice teachers understand the experience of learning to read? (Report No. CS 008 128). Chicago: Annual Meeting of the American Educational Research Association. (ERIC Document Reproductive Service No. ED 260 383).

Iaquintal, A., & Hipskyl, S. (2006). Practical bibliotherapy strategies for the inclusive elementary classroom. *Early Childhood Education, 34*(3), 209–213.

Jimerson, S., Brock, S., & Cowan, K. (2005). Threat assessment: An essential component of a comprehensive safe school program. *Student Counseling,* pp. 11–15.

Jimerson, S., Brock, S., Grief, J. L., & Cowan, K. (2004). Threat assessment: A primer for educators. In *Helping children at home and at school II: Handouts for families and educators*. National Association of School Psychologists.

Johnson, A., & McClam, E. (2014, January 14). 12-year-old in custody after 2 students shot at New Mexico middle school. NBCNews. Retrieved online September 2, 2014, at http://usnews.nbcnews.com/_news/2014/01/14/22302153-12-year-old-in-custody-after-2-students-shot-at-new-mexico-middle-school

Johnson, D. (2013). Power up! The new school library. *Educational Leadership, 71*(2), 84–85.

Jones, B. (2013). Gun violence, videogames, and libraries. *American Libraries, 44*(5), 15.

Jones, R. (1996). Fire! *American School Board Journal, 183*, 24–28.

Kahn, M. (2012). *Disaster response and planning for libraries*. Chicago: Neal-Schumann.

Kamp, D. (2013, December). The day before. *Vanity Fair*, pp. 146–156, 200–201.

Kautzman, A. (2011). Active shooter in the library: How to plan for, prevent, and survive the worst. *Library Leadership & Management, 25*(1), 1–9.

Keyes, J. M., & Stoddard Keyes, E. (2014). *I Love You Guys Foundation* website. Retrieved online March 22, 2014, at http://iloveuguys.org/index.html

Kilmer, R. P., Gil-Rivas, V., & Hardy, S. J. (2012). Students responding to natural disaster and terrorism. *Supporting and Educating Traumatized Students: A Guide for School-Based Professionals*. Oxford: Oxford University Press, 229–45.

Krug, E. G., Dahlberg, L. L., Mercy, J. A., Zwi, A. B., & Lozano, R. (2002). *World Report on Violence and Health*. Geneva: World Health Organization.

La Greca, A. M. (2002). *Helping children cope with disasters and terrorism*. Washington, D.C.: American Psychological Association.

La Greca, A. M. (2008). Interventions for posttraumatic stress in children and adolescents following natural disasters and acts of terrorism. In M. C. Roberts, D. Elkin, & R. Steele (Eds.), *Handbook of evidence-based therapies for children and adolescents*. New York: Springer.

La Greca, A. M., Sevin, S. W., & Sevin, E. L. (2005). *After the storm: A guide to help children cope with the psychological effects of a hurricane*. Coral Gables, FL: 7-Dippity-INC.

Lanier, W. (2014). *Avalanches*. Minneapolis, MN: ABDO Publishing.

Larson, J., & Hoover, J. (2012). Quality books about bullying in the young adult tradition. *Reclaiming Children & Youth, 21*(1), 49–55.

Lee, O. (1999). Science knowledge, world views, and information sources in social and cultural contexts: Making sense after a natural disaster. *American Educational Research Journal, 36*, 187–219.

Libraries and kids: quotes you can use. American Library Association. Retrieved online August 28, 2014, at http://www.ala.org/alsc/issuesadv/kidscampaign/kidsquotes

Loertscher, D. (2010). Shifting our visions for our futures: Leadership as a foundational element for teacher-librarians. In S. Coatney (Ed.), *The many faces of school library leadership*. Santa Barbara, CA: Libraries Unlimited, 133–145.

Lovelace, T. L., & Fridell, M. E. (2008). *Helping children cope with disasters: Initiating dialogue with children's literature*. Northwest Missouri State University, National Social Science Association. Retrieved online at http://www.nssa.us/journals/2009-32-2/2009-32-2-13.htm

Lyle, K. (2013). *Columbine survivor: No child should worry about gunmen*. CNN. April 19, 2013. Retrieved online May 26, 2014, at http://www.cnn.com/2013/04/19/opinion/lyles-columbine/index.html?iid=article_sidebar

Mader, B. (2012, August 2). 2012 drought affects on children's health. *National Children's Health Examiner*. Retrieved online November 23, 2013, at http://www.examiner.com/article/2012-drought-effects-on-children

Manifold, M. C. (2007). The healing picture book: An aesthetic of sorrow. *Teacher Librarian*, (34)3, 20–26.

Marcus, J. (2013, June 7). *Fighting back*. TES. Retrieved online December 27, 2013 at http://www.tes.co.uk/article.aspx?storycode=6338509

Maslow. A. (2013). *Hierarchy of needs: A theory of human motivation*. Amazon Digital Services.

Matthews, G., & Feathers, J. (Eds.). (2003). *Disaster management for libraries and archives*. London: Ashgate.

Matthews, J. (2014, April 18). *Fifteen years after Columbine: What we still need to do to protect our kids*. Fox News. Retrieved online April 23, 2014, at http://www.foxnews.com/opinion/2014/04/17/fifteen-years-after-columbine-what-still-need-to-do-to-protect-our-kids/

McCann, J., & Shand, B. (1995). *Surviving natural disasters: How to prepare for earthquakes, hurricanes, tornadoes, floods, wildfires, thunderstorms, blizzards, tsunamis, volcanic eruptions, and other calamities*. London: Dimi.

McClough, J. (Ed.). (2006). *Caring for kids after trauma, disaster, and death: A guide for parents and professionals*. New York University Child Study Center. Retrieved December 27, 2013, at http://www.aboutourkids.org/files/articles/crisis_guide02.pdf

Media and violence: An analysis of current research. (2014). *A Common Sense Media Research Brief.* Common Sense Media. Retrieved online September 2, 2014, at https://www.commonsensemedia.org/research/media-and-violence-an-analysis-of-current-research

Miller, C., Zickuhr, K., Rainie, L., & Purcell, K. (2013). *Parents, children, libraries, and reading.* Pew Internet. Retrieved online January 5, 2014, at http://libraries.pewinternet.org/2013/05/01/parents-children-libraries-and-reading/

Miller, D. A. (2009). *Guns and violence.* New York: Greenhaven.

Mitchell, M., & Brendtro, L. (2013, Fall). Victories over violence: The quest for safe schools and communities. *Reclaiming Children & Youth* (22)3, 5–11.

Molnar, M. (2013). Districts invest in new measures to boost security. *Education Week, 33*(5), 1–19.

Morawski, C. M. (1997, January/February). A role for bibliotherapy in teacher education. *Reading Horizons, 37,* 243–259.

Morris, R. (2013). Library support for students facing tough times: Resources and stories. *School Library Monthly, 30* (1), 17–19.

Murphy, Sean. (2014, May 19). Fear lingers for students who faced deadly Moore, Oklahoma tornado. *Huffington Post.* Retrieved online May 19, 2014, at http://www.huffingtonpost.com/2014/05/19/moore-oklahoma-students-fear_n_5353879.html?utm_hp_ref=green

Myers, Amanda Lee. (2011a, November 5). Arizona dust storm causes pileups on major interstate. *Huffington Post.* Retrieved online November 24, 2014, at http://www.huffingtonpost.com/2011/10/05/arizona-dust-storm-pileups_n_995793.html

Myers, Amanda Lee. (2011b, December 5). massive phoenix dust storm drastically reduces visibility. *Huffington Post.* Retrieved online December 1, 2014, at http://www.huffingtonpost.com/2011/10/05/arizona-dust-storm-pileups_n_995793.html

Narratives from students at the center. (2006). *High School Journal, 90*(2), 30–50.

National Association of Education of Young Children. (2001). Helping young children in frightening times. *Young Children,* 56(6), 6–7.

National Institutes of Health (NIH). (2013). *Department of Health and Human Services, grant application.* Retrieved online January 23, 2014, at http://grants.nih.gov/grants/guide/pa-files/PA-13-363.html

National Weather Service. (2013). *Office of Climate, Water, and Weather Statistics.* Retrieved online December 15, 2013, at http://www.nws.noaa.gov/om/hazstats.shtml

Neiman, S., & Hill, M. R. (2011). *Crime, violence, discipline, and safety in US public schools: Findings from the school survey on crime and safety, 2009–10.* National Center on Educational Statistics. Retrieved online February 2, 2014, at http://nces.ed.gov/pubs2011/2011320.pdf

O'Brien, Ann. (2011, October 5). Bullying prevention: 5 tips for teachers, principals, and parents. *Edutopia.* Retrieved online on September 2, 2014, at http://www.edutopia.org/blog/bullying-prevention-tips-teachers-parents-anne-obrien

Olweus, D. (1994). Bullying at school: Basic facts and effects of a school-based intervention program. *Journal of Child Psychology & Psychiatry, 35,* 1171–1190.

Olweus, D. (2003). A profile of bullying at school. *Educational Leadership, 60*(6), 12–17.

Ontario Montclair School District. (2014). *Emergency preparedness.* Retrieved online February 20, 2014, at http://omsd.omsd.k12.ca.us/departments/cwa/health/emergencyprep/pages/default.aspx

Ortiz, H. (2011). Educational facilities vulnerability/hazards assessment checklist. *American Clearinghouse of Educational Facilities.* U.S. Department of Education Retrieved online December 14, 2013, at http://www.acefacilities.org/Resources/documents/ACEF%20-Vulnerability-Hazard%20Assessment%20Checklist.pdf

O'Toole, M. E. (2008b). *The school shooter: A threat assessment perspective.* Department of Justice and Federal Bureau of Investigation. Retrieved online September 2, 2014, at http://www.fbi.gov/stats-services/publications/school-shooter

Owings, Lisa. (2014). *The Newtown School Shootings.* Minneapolis: ABDO.

Pardeck, J., & Pardeck, J. A. (1993). *Bibliotherapy: A clinical approach to helping children.* Gordon and Breach [online publishers].

Pearle, L. (2012). When disaster strikes: The Hackley School fire. *School Library Monthly, 28*(4), 8–10.

Peck, Adam. (2014, January 23). there has been an average of one school shooting every other school day so far this year. *New York Times.* Retrieved online January 23, 2014, at http://thinkprogress.org/justice/2014/01/23/3192911/normal-school-shootings/

Pepler, D. J., Craig, W. M., Ziegler, S., & Charach, A. (1994). An evaluation of an anti-bullying intervention in Toronto schools. *Canadian Journal of Community Mental Health, 13,* 95–110.

Peppas, L. (2014). *Hurricane Sandy.* New York: Crabtree.

Petal, M., & Green, R. (2009). The Risk RED School Disaster Response Drill Model and Templates. *The Great California Shakeout.* Retrieved online September 2, 2014 at http://www.shakeout.org/downloads/SchoolDrillsModelandTemplates.pdf

Petersen, G. J. (1996). *The enemy within: A national study on school violence and prevention.* Paper presented at the 76th annual meeting of the Association of Teacher Educators, St. Louis, MO, February 24–28, 1996. ED394907

Petty, K. (2012). Using books to foster resilience in young children. *Texas Child Care Quarterly, 36*(2).

Policy statement: Disaster planning for schools. (2008). *Pediatrics, 122*(4), 895–901. Reaffirmed September 2011. Retrieved online December 14, 2013, at http://pediatrics.aappublications.org/content/122/4/895.abstract

Pollack, S. L. (2006, November). Counseling roles in dealing with bullies and their LGBT victims. *Middle School Journal,* pp. 29–34.

Pollack, W. (2008). Prior knowledge of potential school-based violence: Information students learn may prevent a targeted attack. U.S. Secret Service and U.S. Department of Education. Retrieved online September 2, 2014, at http://www.secretservice.gov/ntac/bystander_study.pdf

Prater, M., Johnston, M. L., & Dyches, T. (2006). Using children's books as bibliotherapy for at-risk students: A guide for teachers. *Preventing School Failure, 50*(4), 5–13.

Raine, A. (2013). *The anatomy of violence.* New York: Pantheon.

Rainie, L. (2014, January 24) *10 facts about Americans and public libraries.* Pew Research Foundation. Retrieved online January 24, 2014, at http://www.pewresearch.org/fact-tank/2014/01/24/10-facts-about-americans-and-public-libraries/

Regan, M. (2014). A false sense of security: Managing the aftermath of a crisis is what the author calls a "new normal" for school communities. *Education Digest, 79*(5), 51–55.

Riedel, C. (2014, January 5). *Missouri town devastated in 2011 tornado opens three new schools, all with safe rooms.* NBC News. Retrieved online January 5, 2014, at http://usnews.nbcnews.com/_news/2014/01/05/22192251-missouri-town-devastated-in-2011-tornado-opens-three-new-schools-all-with-safe-rooms?lite

Robbins, M. (2013). Games and violence. *Library Journal, 138*(5), 88.

Robertson, G. (2013). Survival by committee. *Feliciter: Canadian Library Association, 2*(59), 30–34.

Robertson, G. (2014). Emergency management training for your library: The joys of tabletopping. *Feliciter: Canadian Library Association, 1*(60), 30–32.

Rochmann, B. (2012, December 18). Why it's so hard to help keep kids safe. *Time: Family Matters.* Retrieved online December 23, 2013, at http://healthland.time.com/2012/12/18/school-security-why-its-so-hard-to-keep-kids-safe/

Rossen, E., and Hull, R. (2013). *Supporting and educating traumatized students: A guide for school-based professionals.* New York: Oxford University Press.

Rozalski, M., Stewart, A., & Miller, J. (2010). Bibliotherapy: Helping children cope with life's challenges. *Kappa Delta Pi Record, 47*(1), 33–37.

Safe Havens International. (2013, December 3). 7 lessons learned from Sandy Hook. *Campus Safety.* "School Safety." Retrieved online September 2, 2014, at http://safehavensinternational.org/search/7+lessons+learned+from+sandy+hook

Salamone, MaryEllen. (2006, August 15). *The fifth commemoration of the Attacks of September 11, 2001.* New York: The Families of September 11.

Sanchez, R. (2013, December 15). *Lessons of Columbine and other school shootings helped in Arapahoe.* CNN. Retrieved online December 15, 2013, at http://www.cnn.com/2013/12/14/us/colorado-school-shooting-response/

Sandy Hook investigation. (2013, December 1). *American School & University, 86*(4), 42.

Schools failing when it comes to bullying, violence prevention. (2009). University of Michigan Health System., Ann Arbor, MI: Retrieved online January 21, 2014, at http://www.uofmhealth.org/news/1283schools-failing-at-stopping-bullying

Silver, M. (2013). *Hurricane Sandy: What have we learned?* Centers for Disease Control and Prevention. Retrieved online December 20, 2013, at http://blogs.cdc.gov/publichealthmatters/2013/05/hurricane-sandy-what-have-we-learned/

Simmons, K. (2009, October). False alarms, tornado warnings, and tornado casualties. *Weather, Climate, and Society, 1,* 38–53.

Smith, C. A. (2005, January). *First steps to mighty hearts: The origins of courage.* National Association for the Education of Young Children. Beyond the Journal: Young Children on the Web. Retrieved online November 29, 2013 at http://journal.naeyc.org/btj/200501/smith.pdf

Soulliere, Jessica. (2009). Schools failing when it comes to bullying. *Record Online,* University of Michigan.

Spelling, M. (2008). *Guide to school vulnerability assessment.* U.S. Department of Education Office of Safe and Drug-free Schools.

Spencer, J., & Ashley, W. S. (2010, October 6). *Avalanche fatalities in the western United States: A comparison of three databases.* Springer Science and Business Media. Published online October 23, 2010. Retrieved online January 3, 2014, at http://chubasco.niu.edu/pubs/Spencer%20and%20Ashley%202011%20NH.pdf

Stockton Unified School District Emergency Response Crisis Management, 2009–2010. Retrieved online September 2, 2014, at http://rems.ed.gov/docs/repository/rems_000048_0011.pdf

Stripling, B. (2012, January). Fostering empathy: Who cares? *School Library Monthly, 28*(4) 20–22.

Statement by U.S. Secretary of Education Arne Duncan to Teachers, Principals, and School Administrators in the Aftermath of the Sandy Hook Elementary School Shootings. U. S. Dept of Education website. Dec. 19, 2012. Retrieved at http://www.ed.gov/news/speeches/statement-us-secretary-education-arne-duncan-teachers-principals-and-school-administrators

Suebsaeng, A., Pan, D., & Aronsen, G. (2012, December 14). What happened in the Newtown school shooting? *Mother Jones.*

Swift, J. (2013, December). One year later: Lessons from Sandy Hook. *Connecticut Magazine.* Retrieved online December 20, 2013, at http://www.connecticutmag.com/Connecticut-Magazine/December-2013/One-Year-Later-Lessons-from-Sandy-Hook/index.php?cparticle=2&siarticle=1#artanc

Tan, Shaun. (2011). "The Red Leaf" *Lost and Found.* New York: A Levine.

Tenorio, G. (2013, December 12). Alert youngsters keep school from suffering fire damage. *Poway Patch* (Poway, CA).

Tremper, B. (2014). *Avalanche victim demographics.* Utah Avalanche Forecast Center. Retrieved online November 29, 2013, at http://www.sunrockice.com/docs/us_avalanche_victim_demographics.pdf

Trump, K. (2011). *Proactive school security and emergency preparedness planning.* Thousand Oaks, CA: Corwin.

Trump, K. (2014). *National school safety and security services.* Retrieved online January 31, 2014, at http://www.schoolsecurity.org/about-us/kenneth-s-trump-school-safety-expert/

Twemlow, S. W. M., Fonagy, P., Sacco, F. C., Gies, M. L., Evans, R., & Ewbank, R. (2001). Creating a peaceful school learning environment: A controlled study of an elementary school intervention to reduce violence. *American Journal of Psychiatry, 158,* 808–810.

Vaandering, D. D. (2013). Student, teacher, and administrator perspectives on harm: Implications for implementing safe and caring school initiatives. *Review of Education, Pedagogy, and Cultural Studies, 35*(4), 298–318.

Valenza, J. K., & Johnson, D. (2009, October). Things that keep us up at night: the future of school libraries. *School Library Journal.* 29–32.

VanMeter, V. (1991, Summer). Sensitive materials in U. S. public schools. *School Library Media Quarterly,* pp. 223–227.

Vann, D. (2009, February 12). Portrait of the shooter as a young man. *Esquire.* Retrieved online September 2, 2014, at http://www.esquire.com/features/steven-kazmierczak-0808

Vossekuil, B., Fein, R., Reddy, M., Borum, R. & Modzeleski, W. (2002). *The final report and findings of the Safe School Initiative: Implications for the prevention of school attacks in the United States.* Washington, D.C.: U.S. Secret Service and U.S. Department of Education.

Walker, E. S. (Ed.). (2012). *Children and disasters: Preparedness, response and recovery.* New York: Nova Science.

Washington, I. (2006). Rebuilding a high school library collection after Hurricane Katrina. *Public Library Quarterly, 25* (3/4), 159–178.

Waters, R. (2012, December 17). Gun violence: how research on an American health crisis has been suppressed. *Forbes.* Retrieved online September 2, 2014, at http://www.forbes.com/sites/robwaters/2012/12/17/gun-violence-americas-secret-health-crisis/

Wellheiser, J., Scott, J., & Barton, J. (2002). *An ounce of prevention: Disaster planning for archives, libraries, and record centers.* Scarecrow.

Whelan, D. (2011, October). The bully in the backpack. *School Library Journal,* pp. 29–33.

Whelan, D. (2004). Vandals firebomb Jewish school library. *School Library Journal, 50*(5), 19.

Williams, R. (2009). Knowing leadership when you see it. *U.S. News & World Report, 146*(10), 45.

Winter, M. (2014, January 17). 2 students shot inside Philadelphia high school. *USAToday.* Retrieved online January 17, 2014, at http://www.usatoday.com/story/news/nation/2014/01/17/philadelphia-school-shooting/4588715/

Wood, F. B. (2008). Grief: Helping children cope. *Young Children, 63*(5), 38–40.

Wood, K. (1994). Hearing voices, telling tales: Finding the power of reading aloud. *Language Arts, 71,* 346–349.

Ya-Ling, L. (2008). Helping children cope: What is bibliotherapy? *Children & Libraries: The Journal of the Association for Library Service to Children, 6*(1), 47–49.

Zaleski, W. (2009, March 12). How best to warn people about imminent disasters. ***In Case of Emergency Blog.*** Retrieved online March 2, 2014, at http://incaseofemergencyblog.com/2009/03/12/how-best-to-warn-people-about-imminent-disasters/

Zeece, P. D. (1998). Books for children: Disaster! *Early Childhood Education Journal, 25*(3), 189–192.

Zeece, P. D. (2002). Creating literature safety zones for young children. *Early Childhood Education Journal, 30*(1), 47–52.

Index

About the Author

CHRISTIE KAALAND, EdD, is professor in the School of Education at Antioch University Seattle, where she designed and currently directs the school library certification program. Prior to joining Antioch, she worked for 11 years as a school librarian. Her published works include ABC-CLIO's *Activism and the School Librarian: Tools for Advocacy and Survival*, and she has also written extensively for *School Library Monthly*. Kaaland works tirelessly to improve funding for school libraries through legislative and school advocacy aimed at providing equitable library access for children in Washington schools.

About the Contributor

WILLIAM LOKEY has been involved in emergency management for over 40 years. He has worked as the Pierce County Emergency Management Director (1986–1997) and Assistant Chief for Special Operations for the California Governor's Office of Emergency Services (1997–1999). He joined FEMA's D.C. office as a Federal Coordinating Officer (FCO) in 1999. In his career he has been a first responder for 24 U.S. disasters around the country, including the 1980 eruption of Mt. St. Helens, the Northridge earthquake, the Oklahoma City bombing, the attack on the World Trade Center on 9/11, and Hurricanes Opal, Isabel, Katrina, and Rita.

www.ingramcontent.com/pod-product-compliance
Lightning Source LLC
Chambersburg PA
CBHW081125180926
51611CB00004B/60